Adam Dant's Political Maps

First published in the United Kingdom
in 2022 by
B. T. Batsford Ltd
43 Great Ormond Street
London
WC1N 3HZ

An imprint of B. T. Batsford Holdings Limited

ISBN 978 1 84994 691 9

A CIP catalogue record for this book is available from the British Library.

10 9 8 7 6 5 4 3 2 1

Reproduction by Mission Productions, Hong Kong
Printed in Singapore

Adam Dant's Political Maps

BATSFORD

CONTENTS

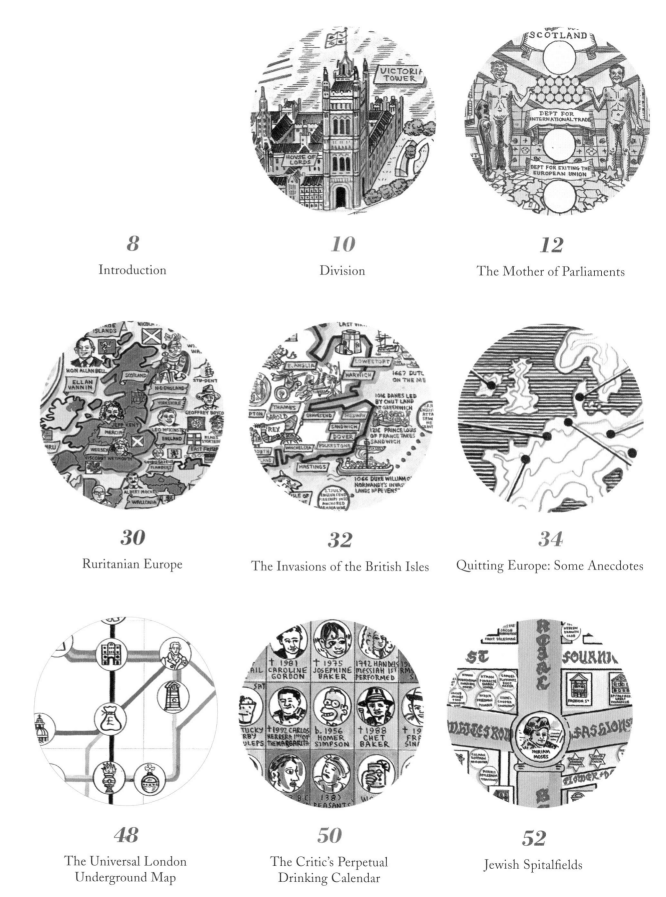

70

Triumph of Debt,
Canary Wharf

72

Moneyscrape

74

Prime Ministers of the
United Kingdom

76

Presidents of the
United States of America

78

Cockney Rhyming America

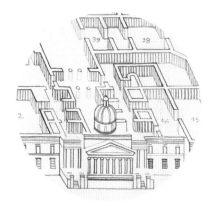

98

A Map of the National Gallery: London's
Collection of Subliminal Images

102

Tulip Fever

104

Coco-opolis

106

Scottish London

108

A Walk Round the End of the World

120

The Meeting of the Old & the
New East End in Redchurch St

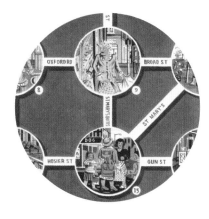

122

Operation Owl Club

124

An Encyclopedy of Ye Age of
Enlightenment Citizens & Kings

126

Stanfords World of
Covent Garden

128

Acknowledgements and Credits

80

New York Tawk

82

Paris Argot

84

The Oxford Meridian

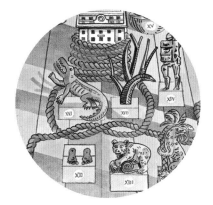

94

Metaphysical Cambridge

96

Theatreland

110

Lindon: A Map of London as it Might Have Been

112

The Mind of East London

114

The Gentle Author's Tour of Spitalfields

116

The Paradise of Sleaze

118

The Great British Beast Chase

THE ART OF POLITICS AND THE POLITICS OF ART: MY LIFE AS A 'POLITICAL ARTIST'

When artists apply the qualification 'political' to describe themselves and their practice it rarely means that they are involved in conventional politics.

Typically the 'political map' as opposed to a 'physical map' is a map that shows territorial borders, and though the maps illustrated and described within the pages of *Adam Dant's Political Maps* do not for the most part describe the world according to its borders, their engagement with 'the political' as a facet of the everyday life of 'the electorate' worldwide can be deemed 'political' enough for the Chinese Communist Party to discourage all the printers of their nation from printing such a book.

Within this particular 'atlas' the parameters of what constitutes a 'political map' are stretched as wide as possible so as to include such diverse subjects as 'A Global History of Chocolate', 'The Metaphysical Life of the Sculptures at the Seats of Learning in Cambridge' and 'The Gutter Language of the Parisian', as well as the more obviously political themes of 'US Presidents', 'British Left Groups' and – for the reader who just happens to also be a serving UK politician – a handy map showing the locations of all the 'Division Bells' in Westminster.

Maps as a tool most famously manipulated by leaders, politicians and the occasional despot have provided us with the current widespread and slightly hackneyed Magritte-style view that 'the map is not the territory'. In the way that the political maps within this tome have been wrangled and wrought to serve all manner of unusual functions, such as in the creation of a 'Cockney Rhyming Slang' glossary of the United States of America, there is barely even a slim possibility of their being taken as depictions of reality.

Rather than existing as vehicles for polemic, proselytizing, pamphleteering or even as mere entertainment, these political maps append the cartographic milieu to an aesthetic and intellectual exercise, creating an armature on which to hang anecdotes, taxonomies, facts and factoids as part of various vaguely philosophical enquiries into the nature of systems, methodologies, social practices and the cataloguing of such seemingly random phenomena into lucid visual histories. Or, to put it more succinctly, in the words of Alan Sillitoe's character Arthur Seaton: 'All I'm out for is a good time … all the rest is propaganda.'

Each map in this volume thus embodies the word 'politics' according to its former meaning, being something for public benefit, at the same time as incorporating the kind of manipulative process for which the word more commonly stands today.

My own engagement as a fine artist with the world of conventional politics really opened up when I was appointed by the Speaker's Committee for Works of Art as the UK Parliament's Official Artist of the General Election 2015. The commission took me the length and breadth of the UK on a tour that involved sketching all manner of campaign events in preparation for the creation of a huge drawing to represent the UK electorate's engagement with the democratic process.

Thus, although the resulting work of art, 'The Government Stable', can be described as a 'political map' of sorts, the political is very much enmeshed in the fairly mundane quotidian world that candidates strive via their manifesto promises to understand, fix and improve. The whole complex work of art is underpinned by a hidden cartographic matrix of the UK's geography and was constructed around the quintessential object of electioneering during the 2015 campaign, namely the 'Ed Stone'.

The Ed Stone, for those who need reminding of the most bizarre highlight of the 2015 general election, was a massive granite monolith inscribed with manifesto pledges that the former Labour leader Ed Miliband had promised to install as a hefty garden ornament at Number 10 had the British people chosen him as their prime minister.

During the 2015 general election campaign the Ed Stone was just one of numerous weird electioneering curios that found their way (hidden in a packing crate in the case of the Ed Stone) into 'The Government Stable'.

A key is provided with the image to allow for the identification of all these objects, as even the most crazed politics geek would have trouble recalling the decorative screen that served as a backdrop for the display of Indian cooking in a Cardiff Balti House by Nick Clegg (then leader of the Liberal Democrats). The unfortunate incident of Ed Miliband and the bacon sarnie, when he'd been caught on camera the previous year awkwardly eating his breakfast at a New Covent Garden Market caff, had led to party press officers asking tabloid photographers to lay down their cameras at lunchtime. A polite but firm Lib Dem lady with a clipboard requested that I too put down my sketchbook while her leader had his curry. I remonstrated that my crayon renderings were not quite in the same milieu as the work of the paparazzi, but she insisted that they were the same thing. 'What if I draw the scene from my imagination on the train home,' I asked. She gave me a stern, humourless look and told me that her team would prefer – and indeed would recommend – that I didn't.

For the most part the various parties' political advisors and fixers of 2015 were charming, helpful and very smart. Things only got tense during unfortunate off-message moments, inevitable cock-ups and perceived biased-media stitch-ups. Until I pointed it out in a sketch, nobody had noticed that the sleeves of the Conservative leader David Cameron's white shirt, which he always wore rolled up ready for the hard job at hand, had actually been steam-ironed thus by one of his loyal and fastidious aides. The SNP's press officer told me he was 'very disappointed' with my illustrated artist's diary entry from a huge anti-Trident rally in Glasgow. When I asked why, he told me that he'd been there with his kids and he hadn't heard any of the crowd shouting to the press pen, 'If any of you are from the BBC we'll come down to London and burn down your houses,' as I'd recorded.

The campaign trail of 2015 was a very friendly one and despite it being the first 'Twitter election' that particular social-media conduit had yet to become the cesspit of hostile opinion and spats that it is today. As a guest and in-house artist on ITN's big election-night broadcast I was even given a biscuit decorated with a blue iced Twitter bird by the company's political pundit. That things change so quickly in the world of politics – the faces, the events, the decisions all so fleeting – for the artist presents a subject that offers both an opportunity to be part of something that seems important and monumental at the time, while also embodying all the transient and, in terms of aesthetically compelling material, a series of extremely mundane-looking elements that attend the creation of legislation.

The current Atelier Dant.

DIVISION

When I was a child, in common with many other young people at the time, I wrote a letter to my local MP asking if they would give me and my classmates a tour of Parliament. My MP wrote back straight away and very soon a bunch of raggedy nine-year-olds were stuffed into the St Laurence's Catholic Primary School's rusty old school minibus and driven from Robert Rhodes James's seat in Cambridge to the seat of power in Westminster.

What I remember in particular from our tour of the upper and lower chambers of Parliament with its dark, scary, wood-panelled, bookcase-lined Victorian corridors was Mr Rhodes James's description of his home in the capital. Did he come here from Cambridge every day? I'd asked. When he informed us that he also had a home in London 'within the division bell' my stupid nine-year-old imagination (and I was not alone in thinking this) immediately imagined our MP having his living quarters inside an actual bell. One of the nuns had been reading to us from *James and the Giant Peach* that term. If a seven-year-old boy could live in a big piece of fruit then why couldn't a politician live in a big bell? It's a good idea.

Once I'd finished drawing this map, which locates several of Westminster's 'off-site' division bells, I realized that in effect our MPs do all live within a big bell.

As well as the bells that sound within the Parliamentary estate to call MPs back to the chamber to walk through the division lobbies whenever there's a vote, there also exists a whole network of extramural tintinnabulary tips to make sure members know that if they don't get back to the Central Lobby within eight minutes then they'll be 'locked out' and miss a vote. When the Houses of Parliament were rebuilt after the destructive fire of 1834, the provision of victuals was not as copious as today, necessitating lunches at the pubs and kitchens of the general environs. The unofficial famous last words of Pitt the Younger were not the prosaic 'Oh, my country, my country,' but 'I could eat one of Bellamy's veal pies,' Bellamy being the chef to the Parliamentary estate.

In creating 'Division: Who Goes Where in Westminster', stories of political 'gourmandizing' are rife. Every venue that has a bell, from the St Stephen's Tavern to the Cinnamon Club, has a colourful tale to tell. The general operation of this system in its off-site incarnation (with 172 locations, according to Wikipedia) is a mystery worthy of investigation by the legendary 'double-0' denizens of the St Ermin's Hotel, whose doormen proudly point out their own recently restored (but possibly 'disconnected') division bell to all visitors.

There's a lot of work that can be done in the pubs and restaurants of Westminster secure in the knowledge that votes will not be missed thanks to this arcane system of wires, circuits and metal cloches. I tried to trace the source of the system but, after being blinded by science and secrets in a basement room behind a door labelled 'Engineering Control' directly underneath Parliament's Central Lobby, I much prefer to stick to the idea that our elected representatives are happily installed and ready for action, owl-like, clustered within a massive bell.

DIVISION

Hand-tinted print
56 × 76cm (22 × 30in)

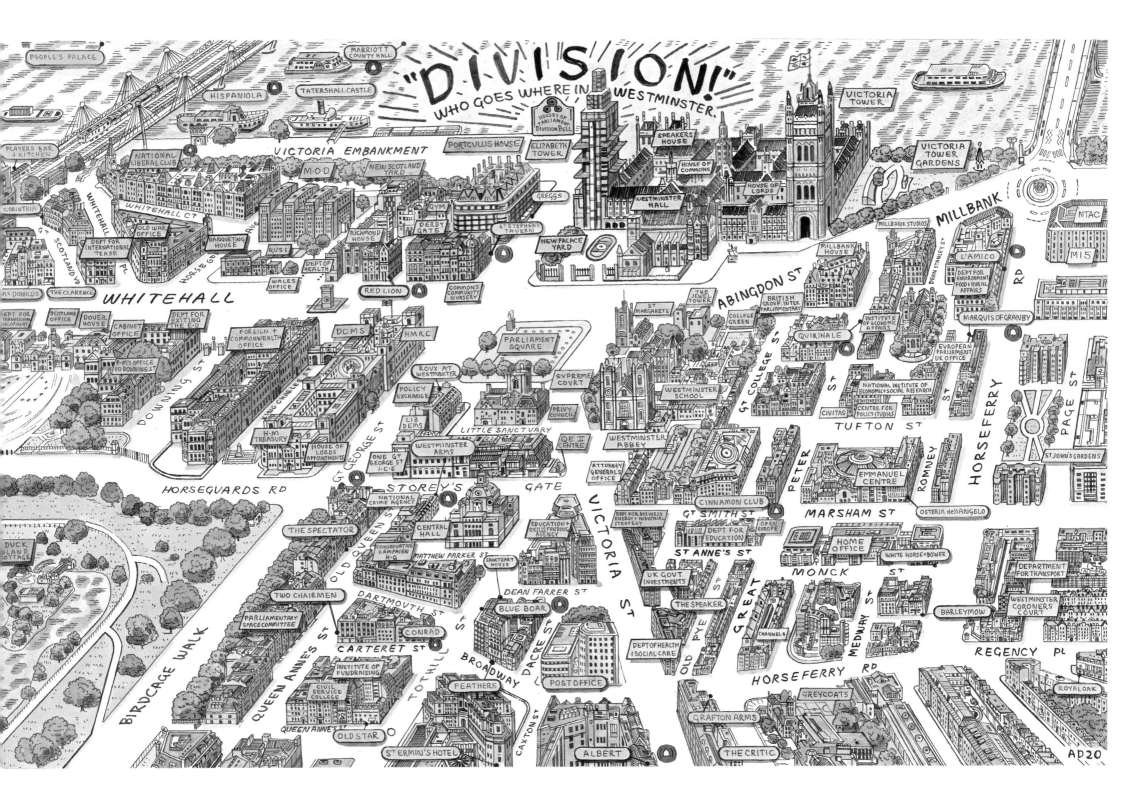

THE MOTHER OF PARLIAMENTS

'The Mother of Parliaments: Annual Division of Revenue' is a contemporary almanac that took its inspiration from the state of British politics in 2017 and the political life of Baron Ferdinand de Rothschild (1839–98), Liberal MP for Aylesbury from 1885 to 1898, who built Waddesdon Manor.

This particular almanac is a modern and subversive response to the exhibition 'Glorious Years: French Calendars from Louis XIV to the Revolution', which showcased a remarkable – but little-known and never before displayed – collection of calendars (originally named 'almanacs'). The exhibition charted the evolution of these calendars from their golden period under the reign of Louis XIV to the French Revolution, when time itself was re-invented.

Our modern politicians are re-imagined through the lens of the official almanacs of the Old Regime. The Mother of Parliaments replaces the French kings and their attributes, with the unusual spectacle of the glorification of modern British MPs, their power and reverence communicated through an everyday print.

Almanacs are political documents, issued as propaganda exercises by the French establishment and, later, by those seeking to overthrow it. As a modern reflection of that spirit, it exists as a print for the British electorate of 2017.

Suitable for hanging on the walls of every British kitchen in place of the ubiquitous kitten calendar or stately-home tea towel, 'The Mother of Parliaments: Annual Division of Revenue' allows for the division of the annual governmental budget among its various departments to be pencilled in from year to year using blank roundels across the design. The central calendar panel lists the birthdays of the UK Parliament's 650 MPs in the manner of a liturgical or Jacobin menologion.

Set against a backdrop of the Central Lobby of the Palace of Westminster is a schematic rendering of the important buildings, figures and symbols of government. Familiar political faces such as the then Prime Minister, the Leader of the Opposition and MPs in charge of the Foreign Office, the Home Office and HM Treasury are rendered in the timeless style of classical allegory.

The depiction of each is elevated stylistically in the manner of the 18th-century French almanac model. The drudgery and grey facade of Civil Service life thus acquires something of the triumphant self-aggrandisement of the French Court or the Revolutionary National Assembly.

ALMANAC FOR THE YEAR 1734 (THE AUGUST PORTRAITS OF THE FIRST BORN SONS OF OUR KINGS)

(Unknown, 1733). Etching and engraving on paper, Waddesdon Manor.

THE MOTHER OF PARLIAMENTS

Hand-tinted print with interchangeable calendar
56 × 76cm (22 × 30in)

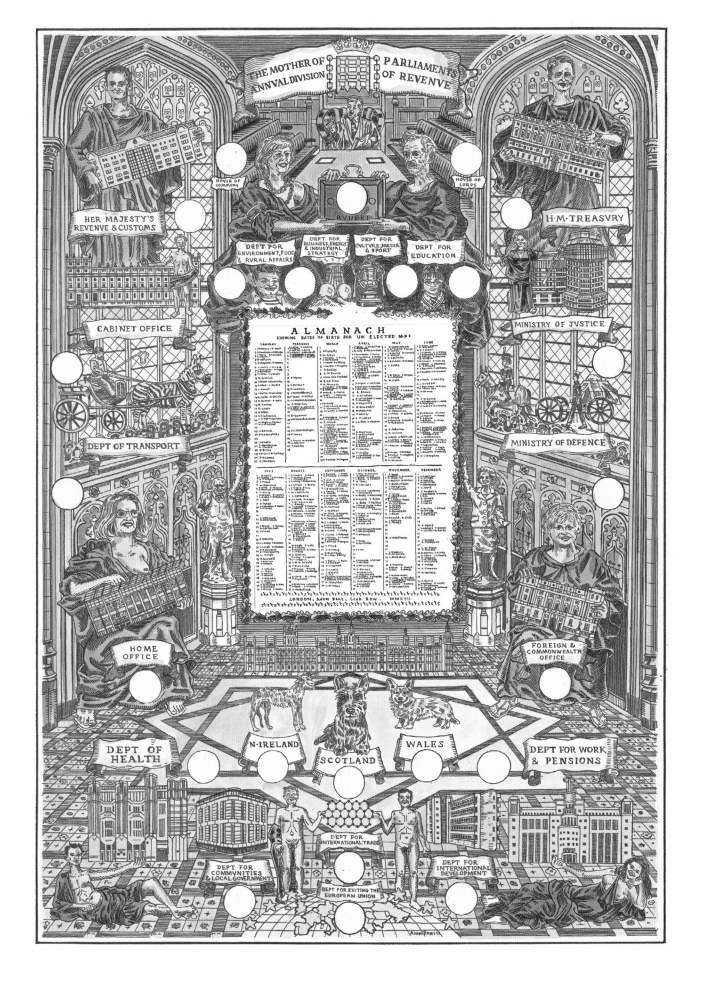

13

The Government Stable

'The Government Stable' currently hangs on the mezzanine at Portcullis House, alongside other works of art from the Parliamentary collection: portraits of MPs styled according to reputation, depictions of cabinets long gone and now the subjects of BBC period dramas, and the occasional image of Parliament on fire in 1834.

The two dozen sketchbooks I had filled with 'lightning' sketches of campaign events, manifesto launches and voter canvassing from across the length and breadth of the UK, politically as well as geographically, have just been donated to and acquired by the Speaker's Advisory Committee on Works of Art. In 2001 this parliamentary committee, under its then chair Tony Banks, had created the role of 'Election Artist'.

Speaking recently to the 2019 Election Artist, the sculptor Nicky Hirst, we both agreed that despite the post being a great privilege for an artist it also had the potential to be a disorientating and nerve-wracking experience fraught with all manner of unfathomable unknowns.

On learning of my own appointment for the 2015 general election there was an expectation from the political candidates, the press and the public that my 'official Election Artist' role might be akin to being some kind of cruise-ship cartoonist or even like a 'War Artist'.

Despite insisting that I was not a caricaturist, John Humphries on the Radio 4 *Today* programme kept asking me who was the best party leader to draw. In frustration I was on the brink of saying that of all the candidates, with his static oration style and 'LibDem standard' looks, Nick Clegg would make the best nude life model. Sadly the Slade's loss is Facebook's gain there.

The war artist comparison was closer, though in my admiration for the 'reportage'-style drawings of Goya, Winslow Homer, Ronald Searle and Linda Kitson I was concerned that a visually dynamic view of an exit poll, for example, or a drawing of Samantha Cameron gloss-painting a park bench might not prove quite as dramatic as the bloody horrors of the Peninsular War.

My initial concerns as to whether an election campaign might provide much in the way of visually interesting material were allayed as soon as I received the first Liberal Democrat Party's 'operational notice' inviting me to a press call at a hedgehog sanctuary.

The campaign trail from one bizarre event to the next continued apace for six weeks. From witnessing the Queen guitarist Brian May handing out enamel prefects' badges with Caroline Lucas under the Brighton bandstand to the DUP candidates hanging from the back of a London bus, *Summer Holiday* style, in the midst of a remote field in Northern Ireland, I soon had a nice sketchbook collection of strange campaign tokens and mementos.

It was the attendant objects, props and effects of electioneering that, for me, when separated from the grip of politics, added a more humanizing and poignant aspect to a political process that can by turns appear cloying, tedious, repetitive, desperate and – in the case of the most recent 2019 election campaign – nasty and divisive.

In the end, politicians are entirely absent from 'The Government Stable'. The closest any of the 2015 contenders come to being depicted is David Cameron, whose disembodied arm, sleeve rolled up (and ironed) is mounted trophy-like on a wall clutching the infamous 'I'm afraid there is no money!' note scribbled by Labour's outgoing Treasury Secretary Liam Byrne in 2010. Nigel Farage is somewhere in there too, but buried deep within a life-size alabaster sculpture of a seething press scrum.

THE GOVERNMENT
STABLE

Sepia ink drawing
2.74 × 2.13m (9 × 7ft)

1. Leeds Town Hall: The Victorian civic architectural splendour of Leeds Town Hall is the venue for the BBC's final leadership orations. Ed Miliband has a slight Madonna-at-the-Brit-Awards stumble from the stage. The ceiling and arches are decorated with the logos of the UK political parties.

2. Central Methodist Hall, Westminster: The clock and pipe organ are from the hall where the BBC's 'Challengers' Debate' takes place. The clock marks the time – 10pm – that polling stations across the UK close and voting ends.

3. Swindon University Technical College water tower and courtyard pavement: Venue for the Conservative Party manifesto launch, the college occupies Swindon's former railway village.

4. Testbed 1 nightclub, Battersea: Hanging from the ceiling are glow-stick lights from the trendy power-cut-hit Liberal Democrat manifesto launch venue. Panels on the ceiling are decorated with the Lib Dem's backdrop of children's handprints.

5. ArcelorMittal Orbit, Queen Elizabeth Olympic Park: The Labour Party election campaign launch takes place in the viewing gallery of the tower. The party leader, introduced by an NHS nurse, enters through a reception line of cheering Labour student activists.

6. Escalators from UKIP's poster on immigration policy.

7. International Climbing Arena, Ratho, Edinburgh: Vertiginous handhold-studded climbing walls provide the backdrop to the Scottish National Party manifesto launch.

8. The White Cliffs of Dover: The United Kingdom Independence Party unveil a campaign poster depicting three escalators travelling up the White Cliffs of Dover at The Coastguard, St Margaret's, with the cliffs, the English Channel and France Télécom on everyone's mobile phones as a backdrop.

9. The National Liberal Club: The announcement of the Lib Dem's public-sector pay rise commitment. Photographers wait on the grand spiral staircase for the party leader to appear. Veterans of election campaign reporting mention that in the past this was how most electioneering took place.

10. Northwood Club, Ramsgate: UKIP's St George's Day celebrations take place with several invited veterans acting as guests of honour at this Kent social club. The club is fully decked out in flags and bunting.

11. Almeida Theatre, Islington: Several celebrities and public figures including Ross Kemp, Baroness Valerie Amos, Emily Berrington, Adjoa Andoh and Mary Creagh speak in support of Ed Miliband's party address on International Development at this decorative Islington venue.

12. Bangor town centre clock tower: Plaid Cymru leader Leanne Wood and supporters have a campaign walkabout in pouring rain through Bangor in support of Arfon candidate Hywel Williams.

13. The Rovers Return, Manchester: Ed Miliband unveils the Labour Party manifesto at the *Coronation Street* soap opera set.

14. Wrightbus factory, Antrim, Northern Ireland: Supporters of the DUP dress in their best party clothes for the manifesto launch at the factory where London's Routemaster buses are made. The number 24 bus to Victoria provides an incongruous backdrop for speeches that start and finish according to the klaxon sounding a break in production-line activity.

15. Sikh temple, Gravesend: David Cameron and his wife visit a traditional gurdwara to celebrate the Sikh Vaisakhi festival. The worshippers have vetoed any 'political' speeches at the temple.

16. Royal United Services Institute, Whitehall: The RUSI was established by the Duke of Wellington in 1831 and provides an impressive venue for a series of talks and a debate on UK defence issues. Attractive chandeliers light the main hall.

17. Sipsmith Gin distillery, Chiswick: Brass distilling apparatus makes a handsome backdrop for a visit to the Sipsmith distillery by London Mayor Boris Johnson and local parliamentary candidate Mary Macleod.

18. Brighton Bandstand: Rock guitarist and pro-badger campaigner Brian May joins local MP Caroline Lucas to promote his 'common decency' political campaign, handing enamel brooches to press and supporters.

19. Thurrock Hotel: UKIP unveil their 'fully costed' manifesto pledges at this Essex motel. The mirror and decorative sconce-lined function room looks over fields of pylons and is filled with unattractive metal-framed fabric-covered chairs.

20. North Laine, Brighton: A bank holiday delays the unveiling of the Green Party's 'Vote Brave' campaign poster. A mural of a cartoon blue whale on the opposite wall serves as a backdrop for numerous TV interviews with party leader Natalie Bennett and local candidate Caroline Lucas.

21. National Grid Training Centre, Newbury: David Cameron delivers a speech on apprenticeships to a circle of mainly young people in front of a huge circuit-breaker in the industrial space. Close by the training centre stands the latest design of pylon as lauded that week by the Pylon Appreciation Society.

22. Lynch Plant Hire, Stanmore: Ranks of heavy machinery and plant are parked behind London Mayor and parliamentary candidate for Uxbridge and South Ruislip, Boris Johnson, as he unveils his Conservative Party 'pledge card' for the capital.

23. DIVEX diving equipment factory, Westhill, Aberdeen: Nick Clegg visits a specialist diving equipment manufacturer with local Lib Dem candidate Sir Robert Smith. The party leader assists a frogman who emerges from a test diving tank in the lobby of the DIVEX HQ.

24. Albert Park Gates, Abingdon, West Oxford: The Liberal Democrat's Battle Bus is launched during a sudden downpour on a quiet residential street in front of a local park. The bus is decorated with designs by art student Liisa Chisholm. Daffodils in party colours planted on the verge side do not fare well under the press and party supporters.

25. Britvic bottling plant, Leeds: George Osborne delivers his 'Northern Powerhouse' speech early in the general election campaign. High vis jacket-clad employees wearing hats and hairnets join a large audience of press in 'The Canyon', a warehouse lined with stacks of bottled mixers, cans of Tango and Fruit Shoots.

26. Thainstone Centre, Inverurie: The noisy sale rings of Thainstone livestock market are the point of arrival for Nick Clegg in the constituency of Gordon, where he lends his support to local candidate Lisa Jardine. Press photographers are unsuccessful in their attempts to persuade the party leader to pose with the goods on sale.

27. Scottish Widows HQ canteen, Edinburgh: Bacon sandwiches, haggis and black pudding are the order of the day as David and Samantha Cameron commence their tour of four countries in one day, beginning in Edinburgh at the insurance company's staff breakfast.

8. Mike Crockart's Lib Dem campaign desk: Most constituency party campaign offices look the same, such as that of the Lib Dem candidate for Edinburgh West. Set in an arcade of shops next to the SNP's local HQ, the operation takes place at a large island of generic mix-and-match office desks strewn with boxes of campaign leaflets, hand-addressed envelopes, pledge cards, coffee cups and – in the case of most Lib Dem campaign HQs – several vases of daffodils.

29. George Square, Glasgow: At the 'Stop Trident' political rally, SNP leader Nicola Sturgeon speaks to a large crowd of various political hues, some of whom have written messages such as 'Bairns not Bombs' in coloured chalks on the pavement.

30. Diane Maclean's 'Carrot Amphora', Stanmore: Boris Johnson commences a walkabout with local Conservative candidate Bob Blackmore up the High Street in Stanmore at the local Sainsbury's supermarket.

31. Pete's Eats, Llanberis: One of the most famous mountaineering hangouts in Britain. The local Plaid Cymru candidate commences canvassing at the foot of Snowdon in the Arfon constituency at this legendary café.

32. Snails Café, Rhiwbina, Cardiff: Ed Miliband buys a jar of chilli jam from the deli counter at Snails during a high-profile visit to the North Cardiff constituency with Labour candidate Julie Morgan.

33. Bird in Hand, Austrey: UKIP candidate for North Warwickshire William Cash is barred from the Bird in Hand after posing outside for his campaign photo with a pint purchased from another pub.

34. Mitchell's Dairy, Inverurie: The SNP candidate for the Gordon constituency Alex Salmond and party leader Nicola Sturgeon greet all their fellow diners on a walkabout through the café before taking tea.

35. Pirate Radio FM: David Cameron visits the studios of the Cornish radio station and uses a 'selfie stick' for the first time.

36. The Trews: Ed Miliband makes a clandestine visit to the apartment of Russell Brand to make an appearance on the actor's online 'Trews' broadcast.

37. Arcola Theatre: The Green Party manifesto is launched at this Hackney arts centre.

38. Hare and Hounds Inn, Hyde: Nick Clegg and Danny Alexander launch a Lib Dem tax cuts campaign poster in pouring rain in a pub car park slightly outside the target constituency of Hazel Grove.

39. The Good Yarn: The UK Monster Raving Loony Party candidate for Uxbridge and South Ruislip, Howling Laud Hope, runs his campaign from the High Street Wetherspoons, who have brewed an ale in his honour and underwritten all his party candidates' deposits.

40. The Coastguard Inn, St Margaret's Bay, Dover: UKIP leader Nigel Farage unveils a poster on immigration policy beside the English Channel in the pub car park.

41. Hope and Anchor, Grimsby: Nigel Farage fails to show up for a fish and chip lunch with a beer named especially for him as his boat trip with reality TV star Joey Essex overruns.

42. London Underground 'Royal Baby' poster: Charlotte Elizabeth Diana, the name of the new royal baby, is announced on social media during a Conservative rally in Bath. The news passes round the audience like a Mexican wave as they await the arrival of David Cameron to the podium.

43. The Painting Pottery Café: The Green Party's 'Vote Big, Vote Brave' poster is unveiled opposite this Brighton café.

44. David and Sam Cameron's steak, ale and stilton pies made by the PM and his wife during a visit to Brains Brewery, Cardiff, during their tour of four countries in one day.

45. Leanne Wood's umbrella, left on the ground during rainy photo-calls on walkabouts in Bethesda and Bangor.

46. St George's Day flags and bunting decorate the Northwood Club in Ramsgate for a UKIP celebration, with several war veterans in attendance as guests of honour.

47. The Northwood Club's Robin Reliant.

48. Portrait of Gladstone from the National Liberal Club, where Nick Clegg announces his party's public-sector pay rise commitment.

49. The Purple Godiva: UKIP candidate for North Warwickshire, William Cash, decks his V8 Jaguar Sovereign out in party livery to use as his campaign vehicle.

50. Samantha Cameron's scooter: A few hours after the announcement of her husband's election victory Samantha Cameron is spotted zipping up Downing Street on a child's micro scooter.

51. Mitchell's Dairy cow: Nicola Sturgeon and Alex Salmond pose with the owner of Mitchell's Dairy while leaning on a life-sized fibreglass cow.

52. Number 24 Routemaster bus to Victoria: The bus provides the backdrop to the launch of the Democratic Unionist Party manifesto.

53. Party campaign vehicles, including the Conservative, Labour and Lib Dem battle buses, the SNP's helicopter and Tessa Jowell's Labour Party 'woman to woman' pink minibus.

54. Leanne Wood's brown leather handbag.

55. A chair from the DUP manifesto launch in the Wrightbus factory car park.

56. Brian May's sweet jar full of 'Common Decency' campaign enamel brooches.

57. The Welsh flag borne by Hywel Williams's standard bearer during walkabouts and canvassing in the North Wales constituency of Arfon.

58. The seating from the UKIP manifesto launch at the Thurrock Hotel.

59. One of Natalie Bennett's pink leather boots.

60. Slogans chalked by SNP supporters on the pavement at George Square, Glasgow.

61. The media scrum that greets the arrival of Nigel Farage, punctuated with cries of 'Nigel, Nigel, Nigel' and 'Lovely day for it, Mr Farage!'

62. Chandeliers of the Royal United Services Institute.

63. Diggers and heavy machinery at Lynch Plant Hire, Stanmore, where Boris Johnson launches his London Conservative pledge card.

64. Placards of protesters waved above the tall stone walls of Hayesfield Girls' School, Bath, during a Conservative Party rally.

65. The new design of electricity 'T' pylon as seen outside the National Grid Training Centre during a visit by David Cameron to promote apprenticeships.

66. A Green Party folding street-campaign leaflet table.

67. The egg that missed David Cameron and his campaign battle bus in Bath.

68. 'Saltire' flags on sale in George Square, Glasgow, during a 'Scrap Trident' rally.

69. Fairy wings worn by a young SNP supporter pictured in one of countless selfies taken by SNP leader Nicola Sturgeon.

70. Decorative screen from the Dabbawalla Indian restaurant, Cardiff, backdrop to Nick Clegg's interviews there as he travelled from Land's End to John o' Groats.

71. Shelf of historic diving gear and objects from the DIVEX factory visited by Nick Clegg on his tour of key Scottish target constituencies.

72. Huge fish at Newlyn Harbour held by Nick Clegg while posing for press photos.

73. David Cameron's lectern.

74. Ed Miliband's lectern.

75. Nick Clegg's lectern.

76. Nigel Farage's lectern.

77. Boris Johnson's lectern.

78. Chiavari designer chairs as used at the SNP manifesto launch and at Nick Clegg's resignation speech.

79. Vince Cable's campaign rucksack and parts of his bicycle, stolen while he painted plates at Richmond Adult Community College with Nick Clegg.

80. The Liberal Democrats' 300th 'stake post' planted outside a house in the constituency of Abingdon and West Oxford.

81. Cake celebrating gay marriage that led to a possible 'conscience clause' in legislation in Northern Ireland becoming an election issue after a bakery refused to bake it.

82. Emergency portable mobile phone chargers as used by all journalists covering the general election campaign in 2015.

83. Liam Byrne's 'I'm afraid there is no money' Treasury letter as brandished during election campaign speeches by David Cameron.

84. 'Tardis' ceramic sign outside a house that Nigel Farage entered while canvassing in Ramsgate, failing to re-emerge until a lengthy time after.

85. Election artist's incongruous chair placed in the middle of the audience seating at the ITV leadership debate broadcast.

86. The Harbourmaster's boat used by Nigel Farage and Joey Essex during a campaign visit to Grimsby.

87. Climbing walls at the Edinburgh International Climbing Arena, location of the SNP manifesto launch.

88. The Exit Poll as seen at the ITN general election news coverage opinion room.

89. Memorial statue of a soldier from the Gordon Highlanders in Inverurie, where Nicola Sturgeon accompanies SNP candidate Alex Salmond on a walkabout.

90. UK Monster Raving Loony Party leader Howling Laud Hope's top hat and megaphone and the wig of party founder Screaming Lord Sutch.

91. Curry cooked by Nick Clegg in the kitchens of the Dabbawalla restaurant in Cardiff.

92. Crossed selfie sticks.

93. Allotment bench painted by Samantha Cameron and Conservative candidate Alec Shelbrooke on a campaign visit to 'The Growing Zone', Kippax.

94. Chocolate hat eaten by Paddy Ashdown to honour his 'hat-eating' remarks made after exit poll release.

PEOPLE

95. Spaniel in a saltire campaigning with Alex Salmond in Inverurie, Scotland.

96. Abseilers assisting broadcasters trail satellite cables from the bottom of the disused quarry pit at the Edinburgh International Climbing Arena up to the TV vans above.

97. Conservative supporters await the arrival of the party leader.

98. Labour supporters await the arrival of the party leader.

99. Donna, the floor manager at ITV's Salford studios, during the televised leadership debate.

100. ITV's floor-swiffer buffs the reflective black stage of *The Voice* immediately before the seven party leaders take to it for the televised debate.

101. Sheep sold at the Thainstone livestock market.

102. Nina Hossein, presenter of *London Tonight*, with Facebook's government and politics expert.

103. Scottish Labour candidate Anne Begg's campaign canvassing team.

104. Female Conservative, Labour and Lib Dem supporters represent the constant direct and indirect canvassing of female voters in key marginal seats as being crucial to the general election result.

105. *Daily Telegraph* political journalist Chris Hope and his daughter are regular fixtures at UKIP campaign events such as on St

George's Day, when seen wearing belled and flagged English jesters' hats during UKIP's history curriculum address.

106. At 15 years old, possibly UKIP's youngest party campaigner.

107. A member of Nigel Farage's canvassing team is accompanied by his guide dog.

108. The landlord of the Northwood Club, host for UKIP's St George's Day celebrations.

109. Jo Coburn, presenter of *Daily Politics*, being 'miked up' by her sound man.

110. Stanley Greene, the legendary photojournalist, is employed by *Channel 4 News* to take photographs during a few days of British electioneering.

111. Guests dressed up for the Democratic Unionist Party manifesto launch.

112. Artist of the Liberal Democrat's battle bus decorations Liisa Chisholm.

113. Nick Clegg's personal security man.

114. Brian May, 'Common Decency' campaigner.

115. Former RUSI director Dr Jonathan Eyal.

116. 'Pillow Boy' Lewis Wilson, who leaped from his window to bounce on a mattress with Alex Salmond during campaigning in Port Elphinstone.

117. 'Milifans' demand selfies when they run into the Labour leader while on their hen night in Chester.

118. Welsh student Labour Party supporters dressed as foxes and badgers tail Leanne Wood around Arfon and Anglesey.

119. Young boy in pyjamas waits up to see Nigel Farage canvassing his parents before bedtime.

120. Vince Cable's canvassing team in Twickenham.

121. UKIP press officer Sarah White.

122. BBC *Newsnight*'s Allegra Stratton and James Clayton.

123. Auctioneer at Thainstone livestock market.

124. London bus driver and fan of Boris Johnson.

125. Workers at the Britvic bottling plant 'Canyon'.

126. *Guardian* cartoonist Steve Bell attends the manifesto launches and televised debates with his huge photographic kit bag to sketch party leaders.

127. Head chef of Dabbawalla Indian restaurant in Cardiff where Nick Clegg cooks curry.

128. *Daily Telegraph* journalist Matthew Holehouse seen in the comfortable seating at the back of the Liberal Democrat's battle bus.

129. Employees/technicians at the National Grid Training Centre.

130. Protestor at Conservative rally in Bath.

131. DIVEX test diver surfaces to meet Nick Clegg.

132. SNP supporter at the party's manifesto launch.

Adam Dant sketching on the campaign trail in 2015.

JOHNSON'S LONDON

W
hen we speak of 'Johnson's London' it is usually to summon up a picture of an unreconstructed male world of journalistic waffle, tittle-tattle and sleaze, where high-minded but puerile puns in ancient Greek are exchanged over gamey luncheons alongside guffawed lascivious and chauvinistic opinions at the kind of London club of whose rowdy life a chap might never tire.

The Johnson of this particular map, however, is not the same Londoner whose life was so assiduously recorded in Boswell's 'Life of Johnson'. The capital depicted cartographically in this 'Johnson's London' is exclusively an arena for the *Blond Ambition* of the subject of biographer Nigel Cawthorne, known simply by Londoners as Boris.

Having served two terms as its mayor, represented its outer boroughs as an MP, haunted all its four corners while climbing the property ladder, given his name to the local bicycles and spent his spare moments creating Routemaster buses from old wine cases, the current (at time of writing) prime minister's life is indelibly that of a Londoner. And, as it is for most Londoners, 'Johnson's London' is studded with significant sites like a topographical memoir.

Johnson's childhood home in Primrose Hill is known as the 'pink rocking horse house' after the toy that was stabled in its bow front window; the north London street where journalist Johnson and a host of other hacks lived was christened 'media gulch'. Other Johnson homes are the sites of more fraught incidents, such as the loud row that ensued after a wine spillage on girlfriend (now wife) Carrie Symonds's white sofa led to complaints from neighbours, or the Colebrooke Row rooftop-shed debacle, which saw Islington Council demand the dismantling of Johnson's jerry-built writing shack after complaints from neighbours, or the curtailing of corridor cricket matches outside his mother's Notting Hill flat after complaints from neighbours.

The map of Johnson's London is flanked by classical columns bearing the portraits of as many relatives on one side as wives and girlfriends on the other. The man himself is depicted as the colossus-like 'world king' of his imagination, his head exchanged for the basketball he famously tossed backwards and into the basket.

The whole pictorial view is overseen by Pericles, a bust of whom is apparently one of Johnson's constant companions. How applicable the ancient Greek statesman's words, 'What you leave behind is not what is engraved in stone monuments, but what is woven into the lives of others,' are to Johnson has yet to be seen. Though the auguries of the neighbours are not entirely favourable.

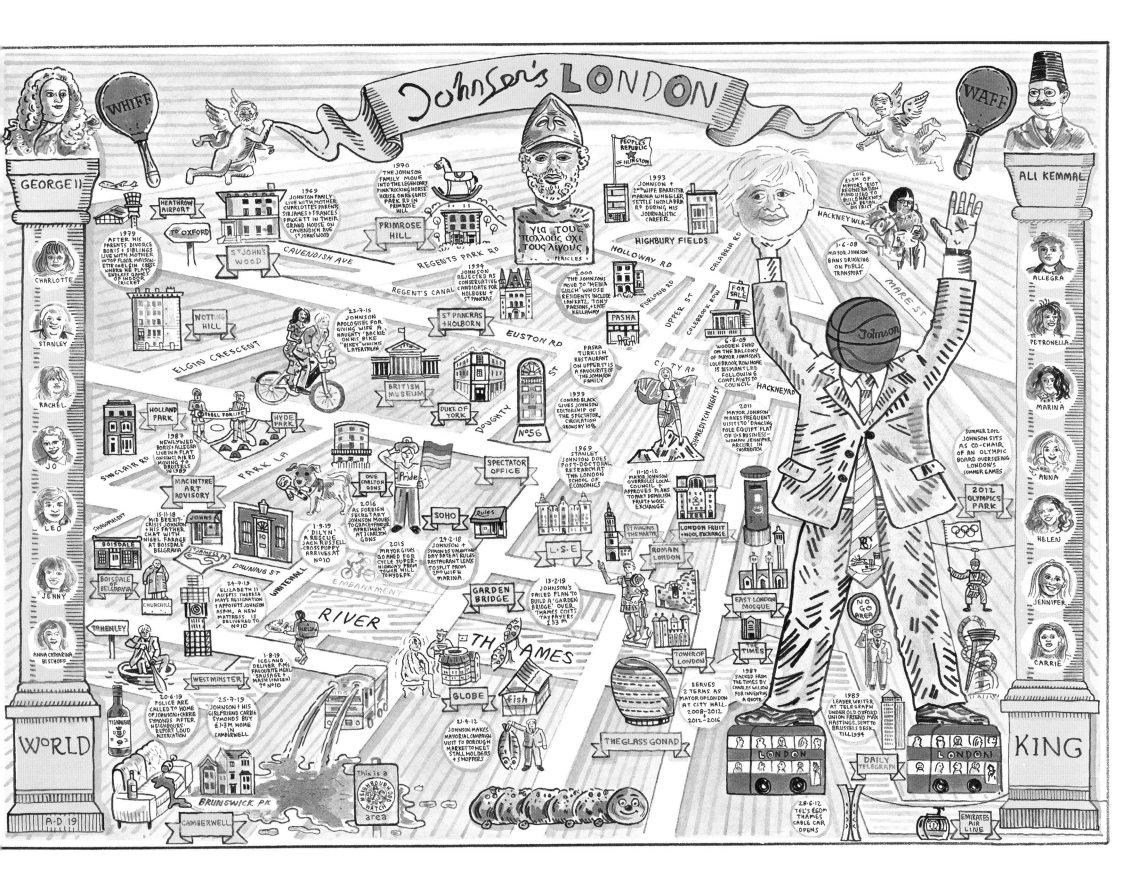

THE MINISTRY + PASSION OF J.C

The role of the political candidate as bringer of salvation and as a wise and moral counsellor has always carried with it unavoidable connotations of and connections to the life and teachings of Jesus Christ. Both are, in effect, seekers of the cross, one on a hill, one on a ballot paper, both apparently between two thieves.

The creation of a timeline of events in the life of the former leader of the Labour Party along the same lines as the other prominent JC was too tempting a tale to resist.

Initial odds of 200/1 certainly made the 2015 election of Jeremy Corbyn as leader of the Labour Party something of a miracle, but the ensuing personality cult, reaching its zenith when he took to the stage to preach to the masses at the Glastonbury Festival, was so messianic in its undercurrents as to be akin to a parody of John Bunyan, or at least bearing the makings of a neat and tidy minor hagiography.

The gospel-like tone of political manifestos in general, and in particular the moral superiority of the Labour Party's project under Jeremy Corbyn, when seen alongside the generally socialist nature of Christian endeavour, addressing the needs of the poor, tackling inequality, etc., serve as a vague backdrop for this satire of 'The Ministry + Passion of J.C' and the colourful (though mostly beige) episodes in the life and times of the man himself as leader.

As an 'epistle to the apostles' (at the time of writing there still exists a devout following who crave a second coming), such satire might appear as akin to mockery and public scourging, thus fitting nicely into the stated trope. At the same time, to detractors, the cultish and questing nature of the Momentum project is also fittingly spoken of along similar lines.

The UK's national press, the City of London and high-end taxpayers in general, aware of the threat that Jeremy Corbyn's economic model posed to the status quo, gained much propagandistic capital from petty stories about the man's life, conduct and foibles. Escapades such as that of 'The Anti-Fascist Beetroot', where a bunch of the eponymous veg was held aloft at a charity event and auctioned to loud chants of 'F*** Fascism', or rumours that the Leader of Her Majesty's Opposition, rather than preparing a considered response to the Chancellor's Budget statement, had instead spent the morning creating decorative patterns from a bag of Medjool dates on his desk, all appeared to be from the realm of fairy tale.

The gross objectification of political candidates as saviours or as conquerors either by supporters or by detractors always ends up looking misguided, overwrought and preposterous when compared in retrospect to either the ensuing horrors and catastrophes or total torpid, mundaneness of life in political office. To the post-modern constituent, images of Adolf Hitler as an armour-clad Teutonic knight on horseback or Napoleon as a Roman Emperor have trampled all the way across the realm of the cautionary to enter the world of the totally stupid. However, politics, like religion, is a continuum contingent on trust and belief. Satire aside, there the similarity should probably end.

THE MINISTRY AND
PASSION OF J.C

Hand-tinted print
56 × 76cm (22 × 30in)

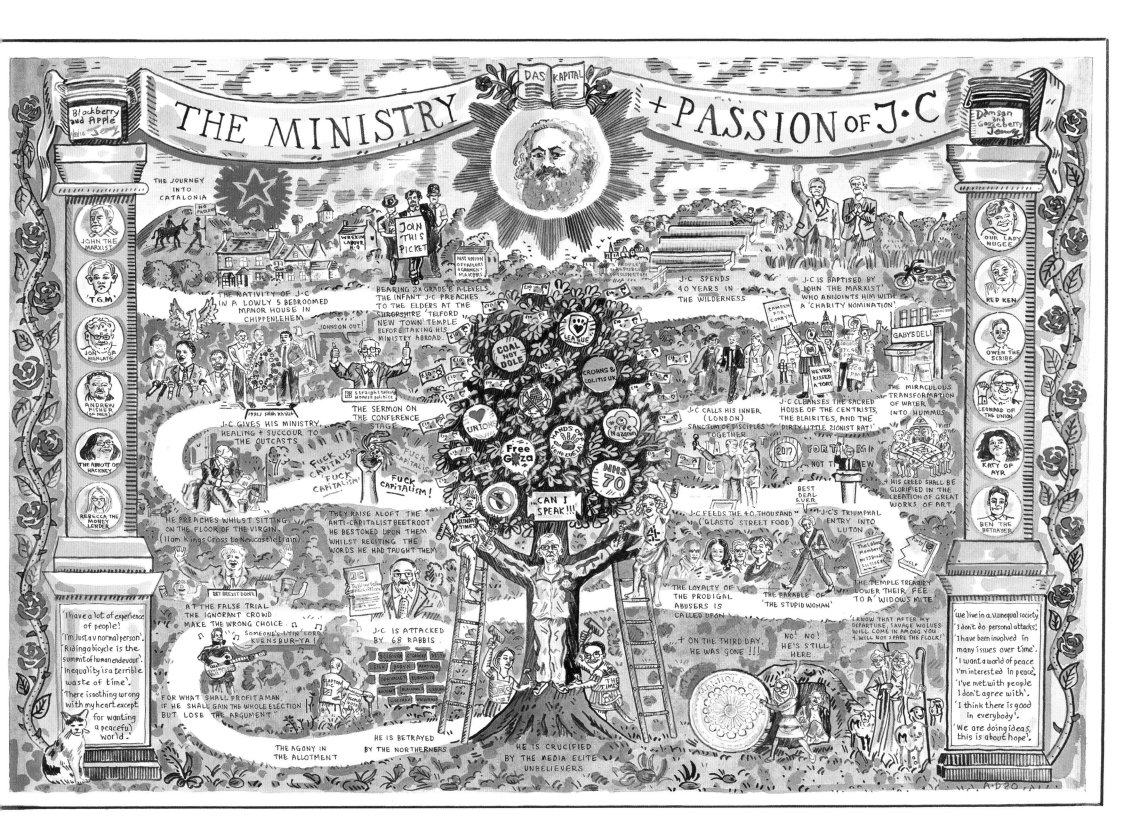

Radical politics, idealistic visions of a fairer future and a zeal for campaigning, protesting, striking and agitating have always been a part of student life, and nowhere more so than in Liverpool during the 1980s, when I was studying at the art school on Hope Street.

Being collared by one particular ex-Cambridge pal who had eschewed his private-school upbringing for a life on the barricades and picket lines of the North meant that the promised 'reunion bevvy' at The Cambridge pub was usually preceded by a two-hour lecture and a discussion on 'anarcho-syndicalism and the dock workers' union' in the backroom of some boozer in the Dingle.

The city was constantly fly-posted with Socialist Worker Party, Revolutionary Socialist Party and Militant Tendency bills, whose big red letters posited shouty interrogations such as 'Nicaragua: What went wrong?' or advanced the causes of Socialism, Marxism, Trotskyism and even Maoism and Stalinism via meetings, marches or calls for direct action. Some evenings out with my radicalized friend and art-school colleagues at the Flying Picket Social Club on Hardman Street really did resemble the 'People's Front of Judea / Splitter!' scene from *Monty Python's Life of Brian*, so complex were the factional issues of the day and the ideologies that came in tow.

At some moments the whole city resembled a kind of biosphere of dissent and – aside from the occasional bus trip to London to 'Smash the Poll Tax' or 'Crush the Alton Bill' – all the idealistic energies of student agitators et al. appeared a bit backwards-looking and out of touch. Events such as the screening of the biopic of Rosa Luxemburg at the Moviola really cranked up the nostalgic outrage, prompting an art-school girlfriend to take up wearing a big floppy beret and a red star for the rest of the term, until a trip to East Berlin swiftly crushed that particular sartorial flourish.

The flow chart of 'British Left Groups' is simple in form, its depictions of the origins of an increasingly fractious lineage of random groups, organizations and parties all originating from a single source, that being a portrait of Karl Marx. Every ideological disagreement, vexatious split and factional divergence, and the resulting parties, groups, tendencies and leagues, is denoted as part of a general chronology. All of these are contained within the body of a giant wooden fish of the type carved by the eponymous hard-luck hero of the Ken Loach movie *I, Daniel Blake*.

As a work of art the image is intended as a historical diversion for the 'political trainspotter' as opposed to being any type of useful 'roadmap to power'.

- All Trades Union Alliance
- Alliance for Workers Liberty (AWL) 1992 (reformed as SOA); Sean Matgamna; de-registered with Electoral Commission 2015 to allow members to join the Labour Party to support Corbyn (Labour Representation Committee (LRC) affiliate)
- Anti-Fascist Action c. 1985–2001
- Anti-Nazi League (ANL) 1977–81; including Chris Williamson
- Anti-Nazi League version 2, 1992
- Association of Communist Workers 1969–77; Harpal Brar; merged with, then expelled from, Socialist Labour Party
- Balham Group expelled from CPGB; Harry Wicks

- Bolshevist-Leninist Faction 1934
- British Marxist-Leninist Organisation 1967; Reg Birch
- British Socialist Party 1911
- Camden Communist Movement 1967
- Campaign for Nuclear Disarmament (CND) 1st phase 1957–63, 2nd phase 1980–83
- Coalition of Resistance 2010
- Communist Action Group 1990s
- Communist Campaign Group 1985; 'anti-revisionist' Mike Hicks
- Communist Federation of Britain (Marxist-Leninist) (CFB (ML) 1969)

- Communist League 1932–36
- Communist League 1990 (NB, different from the Communist League 1988)
- Communist Liaison Group 1991–95; merged into CPB; Andrew Murray, Nick Wright
- Communist Party of Britain (CPB) 1988–present; anti-revisionist, critical of Gorbachev reforms; membership 734 (2017); received 1,229 votes in 2015 general election; Robert Griffiths (General Secretary), Mike Hicks, Andrew Murray (left 2016), Susan Mitchie
- Communist Party of Britain (Marxist-Leninist) 1968

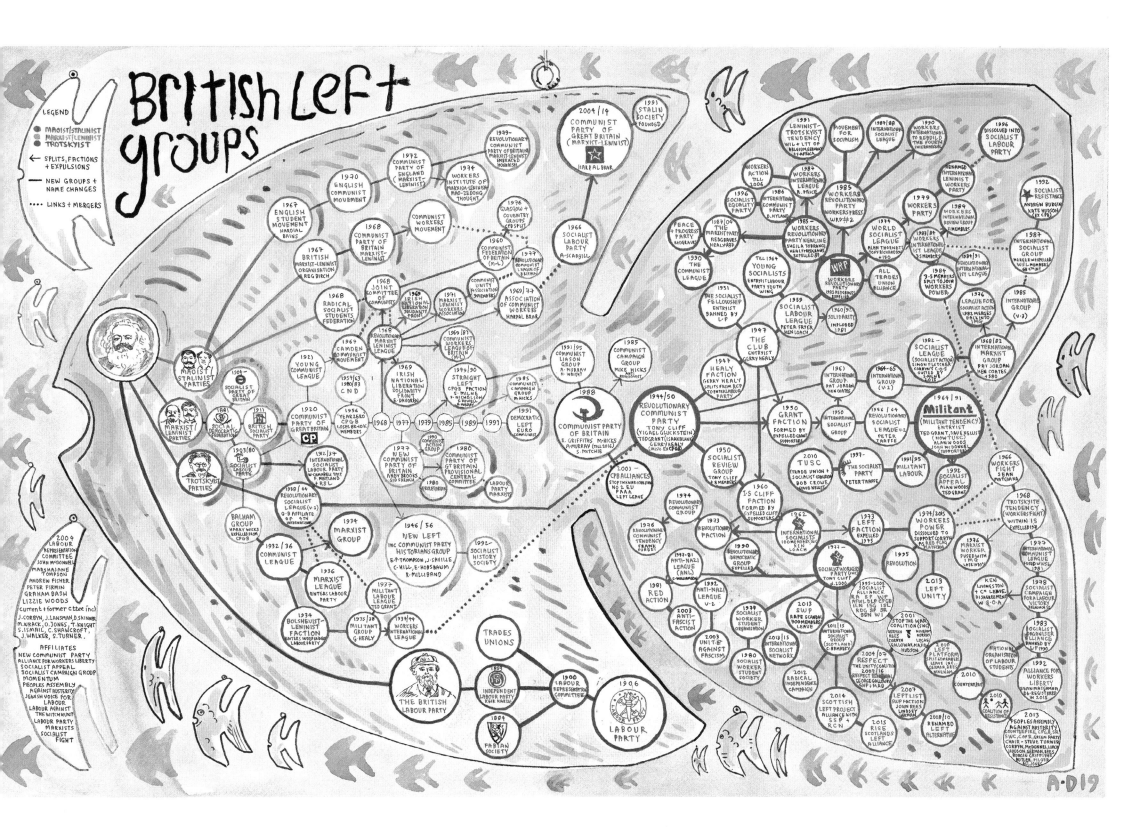

BRITISH LEFT GROUPS

Hand-tinted print
56 × 76cm (22 × 30in)

- Communist Party of England (Marxist-Leninist) 1972
- Communist Party of Great Britain (CPGB) 1920
- Communist Party of Great Britain (Marxist-Leninist) 2004–14; Harpal Brar (chair)
- Communist Party of Great Britain (Provisional Central Committee) 1980–present; plus multiple connections with Trotskyist groups (Socialist Alliance, AWL, Left Unity)
- Communist Unity Association; nine members
- Communist Workers League of Britain (Marxist-Leninist) 1969–81
- Communist Workers Movement 1976
- Counterfire 2010
- CPB alliances (see those groups for details): Stop the War Coalition 2001; No2EU 2009; People's Assembly Against Austerity 2013; Left Leave 2017 (Chair: CPB Gen. Sec. Griffiths)
- Democratic Left 1991; Eurocommunist
- English Communist Movement 1970
- English Student Movement 1967; Hardial Bains
- Glasgow and Coventry groups; split from CFB (M-L) 1976
- International Communist Party 1986; David Hyland
- International Group (version 1) 1961; Pat Jordan and Ken Coates, who left CPGB in 1956
- International Group (version 2) 1964–65; failed to fuse with Revolutionary Socialist League and changed name to …
- International Group (version 3) 1985
- International Leninist Workers Party
- International Marxist Group (IMG) 1968–82; Pat Jordan, Ken Coates; fell from 1,000 to 580 members by 1982, including Ken Loach
- International Socialist Group 1950
- International Socialist Group 1987; merger with expelled members of the Workers Socialist League; became the British section of the Fourth International 1995
- International Socialist Group (Scotland) 2011–15; 39 members at time of split from SWP including Chris Bambery
- International Socialist Labour Party 1912–c. 1937; later Revolutionary Socialist Party 1937–41; William Campbell Tait, Frank Maitland; merged with Revolutionary Socialist League 1938
- International Socialist League c. 1987/88
- International Socialist Network 2013–15; disbands to support Jeremy Corbyn
- International Socialists 1962; membership 100 (1962), 3,000 (1977) including Ken Loach; became the SWP
- International Communist League 1977; fused with Workers Socialist League in 1981, formed by expelled members of the WRP
- Irish National Liberation Solidarity Front 1969; Edward Davoren
- Joint Committee of Communists 1968
- Labour Party Marxists; Stan Keable (expelled from Labour Party in 2017)
- League for Socialist Action 1976; merged back into IMG 1982
- Left Alternative 2008; disbanded 2010
- Left Faction 1973; expelled 1975
- Left List 2007; formed by SWP faction including John Rees and Lindsey German, who fell out with George Galloway
- Left Platform 2010; including Lindsey German, John Rees, Chris Nineham
- Left Unity 2013; Ken Loach, Kate Hudson; membership 2,000 (2014), 1,230 (2016)
- Leninist-Trotskyist Tendency 1991; fusion of WIL, the Leninist-Trotskyist Tendency of Belgium & Germany and a group of South African Trotskyists
- Marxist Group 1934
- Marxist League 1936
- Marxist-Leninist Workers Association 1971
- Marxist Worker 1976; fused with International Marxist Group in late 1970s
- Militant Group 1935–38; Gerry Healy (d. 1989)
- Militant Labour 1991–95
- Militant Labour League 1937; Ted Grant merged with Revolutionary Socialist Party to form RSL 1938
- Militant Tendency (Militant) 1964–91; entryist Ted Grant, Dave Nellist (now TUSC), Alan Woods, John McDonnell (supporter)
- Movement for Socialism
- National Organisation of Labour Students – Socialist Organiser Alliance student section
- New Communist Party of Britain 1977; Sid French, Andy Brooks (LRC affiliate)
- New Left including Communist Party Historians Group (1946–56); E.P. Thompson, John Saville, Christopher Hill, Eric Hobsbawm, Ralph Miliband
- Peace and Progress Party; Vanessa and Corin Redgrave
- People's Assembly Against Austerity 2013; groups include: Counterfire, Communist Party of Britain, Stop the War Coalition, Coalition of Resistance, Green Party, Socialist Resistance; individuals include: Steve Turner (chair), Jeremy Corbyn, John McDonnell, Caroline Lucas, Ken Loach, Kate Hudson, Lindsey German, John Rees, Andrew Burgin, Robert Griffiths (CPB Gen. Sec.), Dawn Butler MP, John Pilger, Bruce Kent, Tariq Ali, Owen Jones
- Proletarian; early 1980s
- Radical Independence Campaign 2012
- Radical Socialist Students Federation 1968
- Red Action 1981
- Respect – The Unity Coalition 2004–07; Respect Renewal 2007–16; George Galloway; unofficially allied to SWP and Muslim Association of Britain (MAB); including Ken Loach
- Revolution 1995
- Revolutionary Communist Group 1974
- Revolutionary Communist League of Britain 1977
- Revolutionary Communist Party 1944–49/50; Tony Cliff (Yigael Gluckstein, born Jerusalem, arrived in UK 1947); 'Ted Grant'

(Isaak Blank, born South Africa, arrived in UK 1934); Gerry Healy (Irish, ex-CPBG)
- Revolutionary Communist Party of Britain Marxist-Leninist 1979–present; rejected Maoism and embraced Albania/Hoxhaism
- Revolutionary Communist Tendency 1976; Frank Furedi ('Frank Richards')
- Revolutionary Democratic Group; expelled early 1990s
- Revolutionary Faction 1973
- Revolutionary Internationalist League 1984–91
- Revolutionary Marxist-Leninist League 1968; Abhimanyu Manchanda
- Revolutionary Marxist-Leninist Communist League 1977–80 (dissolved)
- Revolutionary Socialist League (version 1) (RSL) 1938–1944; British affiliate of the newly formed Fourth International; fell from 300 to less than 20 members by 1944; Fourth International forced a merger
- Revolutionary Socialist League (version 2) 1956–64; Peter Taaffe
- RISE – Scotland's Left Alliance 2015
- Scottish Left Project 2014; alliance with Scottish Socialist Party (SSP) and Republican Communist Network
- Social Democratic Federation 1881
- Socialist Alliance 1999–2005; increasing SWP domination led to resignations; groups participating: Red Action (left 2001), Socialist Party (left 2001), Workers Power (left 2003), Alliance for Workers Liberty, Democratic Labour Party, Communist Party of Great Britain (Provisional Central Committee), Independent Labour Network, International Socialist Group, International Socialist League, Revolutionary Democratic Group, Socialist Perspectives, Socialist Resistance, Socialist Solidarity Network, Workers International
- Socialist Appeal 1992; Ted Grant, Alan Woods
- Socialist Campaign for a Labour Victory 1978; entryist; relaunched 2015 to support Corbyn
- Socialist Equality Party 1996
- Socialist History Society 1992–present
- Socialist Labour League 1959; including Peter Fryer (ex-CPGB), Ken Loach; openly Trotskyist not entryist
- Socialist Labour Party 1903–80; membership 80 (1903), 1,000 (1919); refused to join in foundation of CPGB (1920)
- Socialist Labour Party 1996; Arthur Scargill; membership 385 (2014)
- Socialist League ('Socialist Action') 1982–present; 1980s 500 members within Labour Party; Ken Livingstone, Simon Fletcher (Corbyn's Chief of Staff 2015, ousted by Seamus Milne)
- Socialist Organiser Alliance (SOA) 1983; Sean Matgamna, Mark Serwotka; banned by Labour Party and dissolved 1990
- Socialist Party of Great Britain 1904–present; received 899 votes in 2015 general election
- Socialist Resistance; founders Andrew Burgin (Stop the War), Kate Hudson, (CND, Stop the War, ex-CPB)
- Socialist Review Group 1950; Tony Cliff; eight members
- Socialist Worker Student Organisation 1977; formerly National Organisation of International Socialist Students
- Socialist Worker Student Society, early 1980s; membership declined after 2013 rape scandal
- Socialist Workers Party (SWP) 1977–present; Tony Cliff, leader (died April 2000)
- Solidarity 1960–92; virtually imploded in 1981
- Stalin Society 1991
- Stop The War Coalition 2001; Lindsey German (SWP), John Rees (SWP), Chris Nineham (SWP), Andrew Murray (CPB), Jeremy Corbyn (chair 2001–11, 2015–16), Caroline Lucas (chair 2011–15), George Galloway, Kamal Majid (Stalin Society), Kate Hudson
- Straight Left 1979–early 1990s; faction within CPGB; Fergus Nicolson, Andrew Murray, Seumas Milne, Steve Howell
- The Club 1947; entryist; Gerry Healy
- The Marxist Party 1987–2004; Gerry Healy, the Redgraves
- The Socialist Fellowship; entryist; banned by Labour Party in 1951
- The Socialist Party 1997–present; Gen. Sec. Peter Taaffe
- Trotskyite Tendency (Workers Fight) 1968; within International Socialists; expelled from IS 1971
- TUSC (Trade Union & Socialist Coalition) 2010; Bob Crow (founder), Dave Nellist (leader)
- Unite Against Fascism 2003
- Workers Action; petered out by 2006
- Workers Fight 1966; Sean Matgamna
- Workers Institute of Marxism-Leninism-Mao Zedong Thought 1974
- Workers International League 1937–44
- Workers International League (WIL) 1987; Richard Price
- Workers International Review Group 1984; 11 members
- Workers International to Rebuild the Fourth International 1990
- Workers Internationalist League 1983–84; 35 members
- Workers Party 1979
- Workers Power 1974–2015; including Paul Mason; briefly joined Workers Fight in 1975; dissolved to support Jeremy Corbyn in Labour Party as Red Flag Platform
- Workers Revolutionary Party (WRP); Gerry Healy and the Redgraves (expelled 1985)
- Workers Revolutionary Party Newline (WRP#1) 1985–present; Gerry Healy and the Redgraves (expelled 1987); Sheila Torrance
- Workers Revolutionary Party Workers Press (WRP#2) 1985
- World Socialist League 1974; Alan Thornett, Tony Richardson; maximum 150 members
- Young Communist League 1921; CPGB Youth Movement
- Young Socialists; Labour Party youth wing entryist group; wound up 1964

STOP THAT BREXIT

*D*astardly and Muttley in Their Flying Machines was a popular children's cartoon from the early 1970s, a spin-off from *Wacky Races* that saw the villainous Dick Dastardly and his canine sidekick as First World War flying aces trying to catch a messenger pigeon. As its catchy, repetitive theme song announced, the plot of each episode always had the same single-minded aim: 'Stop That Pigeon'. The increasingly ridiculous and desperate measures employed in attempting to do so differed, but all of them predictably fizzled out or backfired and ended with crashes, collisions and explosions.

'Stop That Brexit' is a novel timeline across which are depicted the various moments during the lead-up to the UK/EU membership referendum and the passage and delivery of the resulting UK Withdrawal Bill, during which period (to name just a few) the governments of the UK, Scotland, Wales and Northern Ireland, the Labour Party, the Green Party, the Liberal Democrats, the SNP, 36 FTSE 100 companies, the Prime Minister, three of his predecessors as well as his successor, the majority of the House of Commons and the House of Lords, the Secretary-General of the United Nations, the Governor of the Bank of England, the President of the USA, the President of China, the World Trade Organization, the International Monetary Fund, NATO, most UK newspapers, the TUC, the heads of the armed forces, a host of celebrities and public figures from the creative industries, Nobel Laureates, university leaders, economists, lawyers, historians, all 20 Premier League football clubs and a red-faced shouty man on St Stephen's Green, Westminster, with a big comedy megaphone all tried to 'Stop That Brexit'.

STOP THAT BREXIT

Hand-tinted print
56 × 76cm (22 × 30in)

RURITANIAN EUROPE

The phrase 'United we stand, divided we fall' might have its origins in the New Testament and live most clearly as a palpable virtue in the post-colonial United States of America, but as a perennially popular chant it probably finds itself most readily utilized in the public political arena when the cause of one group is pitted against that of another – usually bigger – group.

The process of championing difference and distinct identities and campaigning for the rights of self-determination, especially in matters fiscal, can be extended *reductio ad absurdum* until those people over the river, over the road, over the other side of the room, are asked to cede and transfer all controlling powers to the cause.

Thus this map of secessionist – or, pejoratively, 'Ruritanian' – Europe is blanketed with all manner of vexatious peoples united by common but unique causes, all here highlighted in red and illustrated with portraits of their champions and figureheads. Some fights for independence seem straightforward and binary but fraught with violence, such as the fight between Armenia and Azerbaijan over the contested territory of Nagorno-Karabakh. Elsewhere the desires of other secessionist movements, which at first seem similarly clear, when unpicked and their multifarious contingencies and facets drawn out into the open suddenly appear as bizarre, baffling and bonkers

as the story told in the 1949 Ealing comedy *Passport to Pimlico*, in which a small London borough discovers its claim to be a part of Burgundy, swiftly declaring itself exempt from the legal jurisdiction of the British government and consequently from post-war rules on food rationing.

The fictitious Professor Hatton-Jones who verifies this claim is depicted here on the map of 'Ruritanian Europe' alongside very real campaigners for independence such as Nicola Sturgeon of the Scottish National Party. Her anti-unionist project, while seemingly straightforward in intent, is, like the cause of Professor Hatton-Jones, crisscrossed all over with contingencies that when viewed as part of a broader canvas are irreconcilable without the deployment of a complex mess of *Animal Farm*-type legislation, illogical contradictory process and overarching hubris.

The intricacies and complex cultural journeys that underpin all the myriad peoples that coalesce into the *E pluribus unum* of the United States of America are essentially united as well as trumped by a formal set of guiding principles in the shape of the US Constitution. This is something that the equivalent formal tenets of the European project have found harder to circumscribe in the face of uncommon histories, folk traditions, enmities and even sausages.

RURITANIAN EUROPE

Hand-tinted print
56 × 76cm (22 × 30in)

THE INVASIONS OF THE BRITISH ISLES

The British Isles have not always been so difficult to invade. Around 8,500 years ago, when what we now call Britain was still part of the continent of Europe, marauding hordes could just march unimpeded across what was once known as 'Doggerland' to claim whatever they wanted.

However, as it was then part of a contiguous land mass and lacking the serious natural defences (the sea) that make capturing islands such a tempting challenge for would-be invaders in the first place, the land we now know as the British Isles would have been like Switzerland or Poland or France, just another place on the way to somewhere else.

The expansion of the Roman Empire famously halted at the Scottish border where, as represented on this timeline map of invasions of the British Isles, the territory to the north was home to savage, blue-painted cannibal tribes of Picts et al. The remains of Hadrian's Wall suggest not so much that the northwards push was unsuccessful but more likely that the wall was prompted by a desire to keep such uncivilized people out of the empire, the invadee becoming the potential invader.

The founding in around AD 43 of the 'Classis Britannica' fleet, charged with logistically supporting the Roman invasion of Britain, could be argued to mark, in effect, the birth of the Royal Navy, in that once the invasion had proved successful the fleet was used to defend the English Channel.

In stories of the defence of the English crown, numerous outcasts, malcontents and pretenders appear on this map as religious and family conflicts and contested lines of succession saw nations across the Channel hosting would-be usurpers until they were able to muster enough support and firepower to attempt a landing at the coastal home of some like-minded malcontents and launch a full-scale 'failed invasion'.

Discounting the Nazi German invasion of the Channel Isles and a minor skirmish in 1940 on Graveney Marsh that ended with German and English troops enjoying pints of beer in the Sportsman pub, the last successful occupation of Britain took place in 1667, when the Dutch fleet sailed up the Medway. In a deeply embarrassing incident for Charles II, the Dutch set fire to several British ships at Chatham Docks and took control of the Isle of Sheppey for a few days. The French naval force that mounted the last unsuccessful invasion in 1797 surrendered in the face of what they assumed were legions of British troops on the shoreline. The troops turned out to be thousands of local Welsh women of Fishguard dressed in their traditional scarlet tunics and tall black felt hats who had turned out hoping to watch a good scrap.

Invasions are often aided by the perfect set of circumstances but they are by no means always calculated or strategic. The European wolf, the grey squirrel, the beaver, the American crayfish, the rhododendron, Japanese knotweed and the scourge of west London's picnickers the green parakeet, despite often being described as invaders, are mostly mere unwitting migrants in search of suitable new climes carried to the British Isles by chance, in much the same fashion as the vaporizing laser death ray-enabled spaceships of the opportunistic alien invaders of the future.

THE INVASIONS OF
THE BRITISH ISLES

Hand-tinted print
56 × 76cm (22 × 30in)

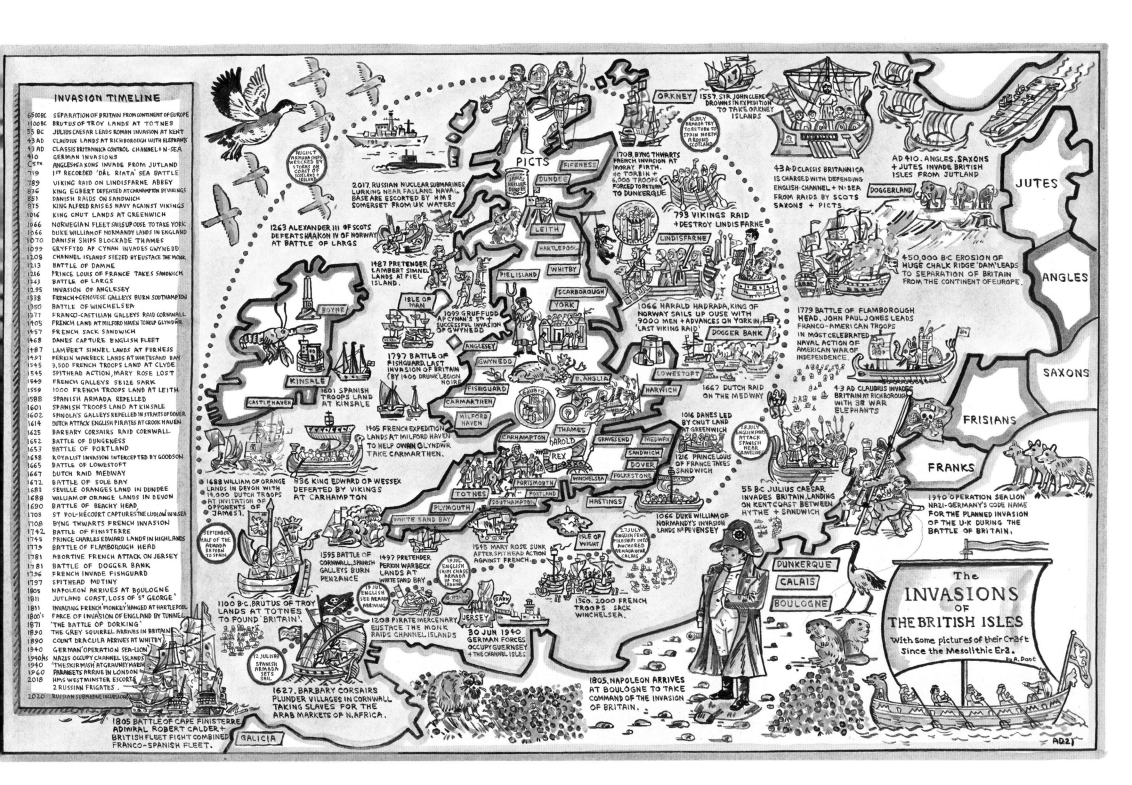

The INVASIONS OF THE BRITISH ISLES

With some pictures of their Craft Since the Mesolithic Era.

By A. Dant

QUITTING EUROPE: SOME ANECDOTES

SWEET AFTON / IRELAND

I didn't know any artists when I was a teenager. My girlfriend kept telling me interesting stories about a friend of her family who was an artist. His name was Ernie, I think he was a Dubliner, he lived somewhere in Yorkshire. Ernie became a hero and a role model to us, even though we only had one of his drawings to study. It was a charcoal portrait of my girlfriend's mother. When Ernie and his wife, who was a nurse, visited us in Cambridge I was quite nervous. At the pub I offered Ernie a cigarette, and he told me that if I wanted to be a Great Artist then I should smoke Sweet Aftons. They were from Dundalk, he told me: 'Flow gently, sweet Afton, among thy green braes, flow gently. I'll sing thee a song in thy praise'. I bought a pack from Colin Lunn, the tobacconist on King's Parade. Lovely cigarettes in a beautiful primrose-yellow box. But really, way too dear for art students who could get ten Dunhills from the machine at the Live and Let Live for 50p.

GAULOISES / FRANCE

After school had finished my girlfriend and I ran away to Paris. I think we were about 16 or 17. We had a hundred quid to last us two weeks; she had an emergency hundred hidden in her shoe, but only told me about it when we were on the ferry home. We bought cigarettes from the Gare Saint-Lazare. The soft blue packet had a picture of what looked like Asterix's hat on it. We smoked them in our room at the Hotel des Fleurs and watched from the window the snaking queue of Arab men outside a brothel. I smoked Gauloises for a few weeks after returning home, until I lit one at my grandma's house. My father ran to open the window, shouting, 'Flaming Nora! What kind of bohemian, foreign compost is that?'

GITANES / FRANCE

When my extremely chic *Parigot* neighbours inevitably moved back to Paris they invited me to stage an exhibition in an ex-fortune-teller's booth they'd discovered for rent up the Passage des Panoramas. They were now both in proper employment. He had found a position at Chanel and had purchased an expensive Barrow and Hepburn briefcase that he carried with him everywhere. A couple of years later, while sharing a Japanese lunch in a very unfashionable businessman's restaurant, I asked him what he kept in the briefcase, which was beside his seat, half-slid under the pink tablecloth. He picked it up, snapped the clasps and opened the lid to reveal that the briefcase was empty except for a carton of 200 Gitanes. 'I couldn't think of anything better to put in it,' he told me.

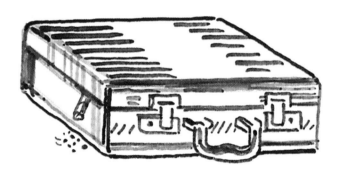

QUITTING EUROPE

Ink drawing
56 × 76cm (22 × 30in)

Ducados / Spain

I thought I'd surprise a girl I really wanted to marry by asking her to come with me to Barcelona as soon as she had landed in London from the States. Princess Diana had been killed just the day before. The Spanish tobacconist went on and on about 'Le Dee-Dee … Le Dee-Dee' as he dabbed his eye, Baptiste-style, with a pretend handkerchief. That was when my wife-to-be spotted the moulded plastic chicken alarm clock in the dusty glass cabinet behind him. 'I want that chicken,' she said. 'I have to have it.' Once the tobacconist showed us what the chicken did, I remember she said 'Sold!' It was the first time I'd heard anyone who wasn't an auctioneer say that.

The chicken alarm clock cock-a-doodle-dooed in the morning, and when one popped down its red plastic comb said, '*Buenos dias.*' There was a slot in its crooked wing to hold a disposable lighter in readiness for one's first cigarette of the day. We were smoking Ducados. The church bells had already woken us.

SG Ventil / Portugal

A lot of British artists had been flown out to Portugal by the British Council for a big exhibition opening at Belém. After the reception, drifting across the city from one event to another, I was fortunate to be befriended by the young Lisboa art gentry. At a local bar, which appeared to consist of nothing more than a makeshift counter and a shelf of bottles and glasses, we smoked the local SG cigarettes, spoke about art and listened to a woman wailing with a guitar in the corner. One of the Portuguese artists told me that during the Salazar era, possession of a cigarette lighter had been a criminal offence. She then went on to tell me how she felt that I possessed a deep sadness inside me. I told her she was mistaken and that it must be the smoke and the music.

MS / Italy

One of the most desperate, hilarious and confounding moments in my life as a smoker came when I realized that the 'MS' on Italian cigarettes did not stand, as locals told me, for '*Merde Secco*', but for '*Monopoli di Stato*', that the workers at the state tobacco warehouses had gone on strike and that the lovely couple who owned the *tabbachi* at the end of my street were rapidly running out of fags. They apologized sincerely and suggested I walk a bit further into Parioli to their friend's shop. Their friend quickly ran out of stock too, and in the morning – after I'd walked for over an hour to a gas station near the ring road where the Mafia demanded 100,000 lira for a pack of Albanian counterfeit Marlboro Lights, then returned to my print workshop to find my studio mate attempting to make roll-ups from ancient, dried-up pipe tobacco – I decided to quit. But not before sending postcards to as many friends and family members as possible inviting them to spend a couple of days in Rome, on the condition that their journey took them via the duty-free shop.

PRINCE / DENMARK

In the midst of the painful and prolonged Italian tobacco strike, when all the shelves of all the *tabbachi* had been completely divested of stock and the price of a black-market packet of Marlboro Lights matched my monthly stipend, I suddenly came up with a brilliant idea. My uncle was a sailor; I remembered him telling me when I was a small child that as well as sailing the Baltic, he often stopped at ports on the Mediterranean and Adriatic. I telephoned my aunt, who told me that my Uncle Bob was just leaving Copenhagen for Italy. Very soon I was in Anzio, in the galley of Bob's ship, with a couple of fellow nicotine-starved artists and George, a teenage archaeologist whose wealthy parents had sent him to dig in Italy during his gap year. We spent the day drinking wine and brandy, listening to my uncle's stories and smoking Danish Prince cigarettes. We headed back to Rome laden down with cartons. When George's Italian odyssey ended and his well-heeled parents came to collect him, he handed me a small parcel. 'This is for you,' he said. 'Don't tell my parents I smoke.' I unwrapped the package to find a fag-ash-lined Colosseum ashtray and a Trevi Fountain lighter.

NIL / AUSTRIA

The chain-smoking regulars in Vienna's coffee houses certainly add plenty of fuggy, old-world 'easy charm' to augment one's lunch in these elegant, civilized spaces. And *beuschel* (a kind of ragout made from heart and lungs) always seems the most appropriate dish to accompany a pack of Austrian Nil cigarettes.

In Vienna I spent most of my days in one coffee house or another, drinking schnapps, chain-smoking, eating lungs and, like many tourists, vainly attempting to enter the lost world of Stefan Zweig, Robert Musil, et al.

I didn't realize that it was normal for patrons to be contacted on the establishment's phone until my Austrian friend called me at the Café Landtmann to let me know that she was on her way with her baby. I told her that it was a bit smoky for a baby, so she asked me to jump on a tram and to meet her at a 'child-friendly' café further round the Ringstrasse.

The child-friendly café where we met was even smokier but most of the tables were occupied by very noisy toddlers eating sausages. This, I imagined, was the real Vienna.

WEST / GERMANY

Just after the Berlin Wall had fallen I made friends with the students at the art school in the former East Berlin. My visits to their studio at Monbijou Park were taken as a kind of betrayal by my colleagues at the art school in the former West Berlin, where I was supposed to be studying, but when I asked them to visit the 'Osties' for a barbecue, their curiosity got the better of them. As the S-Bahn train crossed the infamous 'death strip' we saw a huge trackside billboard showing a picture of a girl offering a cigarette to a swami sitting on a bed of nails. It was an advert for West cigarettes. 'Test the West', the legend read.

My new friends in the East gave us 'bathtub schnapps' at the barbecue. 'Gift!', they said, offering us tumblers. I didn't realize that 'gift' was the German word for 'poison'. Not wanting to refuse a gift, I drank far too much of it. One of my studio-mates offered to give me a lift home on the back of her bicycle. As we pelted past the Tiergarten, my leg clipped a tree and I went flying from the back of the bicycle, and landed on my head in the middle of the pavement. Braking, my friend looked round and shouted, 'That'll teach you for testing the East!'

RUDIMENTARY LONDON

This bold assemblage of pictographic squiggles represents an attempt to visually encapsulate all the nuanced geographic complexity of the capital city but executed in a single minute. Inevitably the resulting graphic ends up looking like a piece of Chinese calligraphy or the artistic output of a Zen monastery. The swiftly rendered brushwork flicks and flourishes when subjected to the woodblock print process to become a 'printed map', also bearing a particular resemblance to 'ink rubbings' such as those produced from 13th-century Suzhou astronomical charts.

But essentially, all these comparisons, random and superficial, are made to stick the idea of 'rudimentary' mapping exercises into the realm of reductive poetics. The globe as a bubble, our lives in a grain of sand, the futility of trying to record all our diverse travails and triumphs in a single flick of the wrist being akin to the flick of the wrist of a vengeful and almighty creator as they sweep all matter aside.

As a depiction of the specific terrain of London it's all there, to some extent or another, according to this particular cartographic conceit. London can immediately be found in that confusing crook in the River Thames, bisected east–west by the Fleet (Farringdon Road), Euston Road skirting its upper edge and splodged with familiar sites and landmarks: Nelson's Column at London's heart, Hyde Park and its Serpentine lake to the west, Regent's Park and its inner circle at the top; there's Tower Bridge and possibly St Paul's (or maybe that's the Central Criminal Court at Old Bailey?). Between the bold, black main thoroughfares the knitted network of London's back alleys is formed from the inky infill of chiselled gouges in the woodblock.

That we are drawn into a conversation of visual transference, like those we have when describing familiar shapes seen in the clouds, is akin to the imposition of the human order of cities on what are essentially convenient aspects of the random terrain of geomorphology.

RUDIMENTARY LONDON

Woodcut print
56 × 76cm (22 × 30in)

39

THE LONDON REBUS

There are 57 pictographic clues on 'The London Rebus' and you don't even need to know how to read or anything about the capital to solve them.

Applying this ancient form of puzzle to the map of London, the city becomes a beguiling collection of pictographs. Were it not for the River Thames going through the image, London would be practically unrecognizable, appearing to be nothing more than a strange lattice of Egyptian hieroglyphs. However, the presence of the City starts to appear as the viewer slowly (or quickly) deciphers these visual clues revealing the names of London's familiar landmarks and neighbourhoods.

THE LONDON REBUS

Hand-tinted print
56 × 76cm (22 × 30in)
Answers on page 128

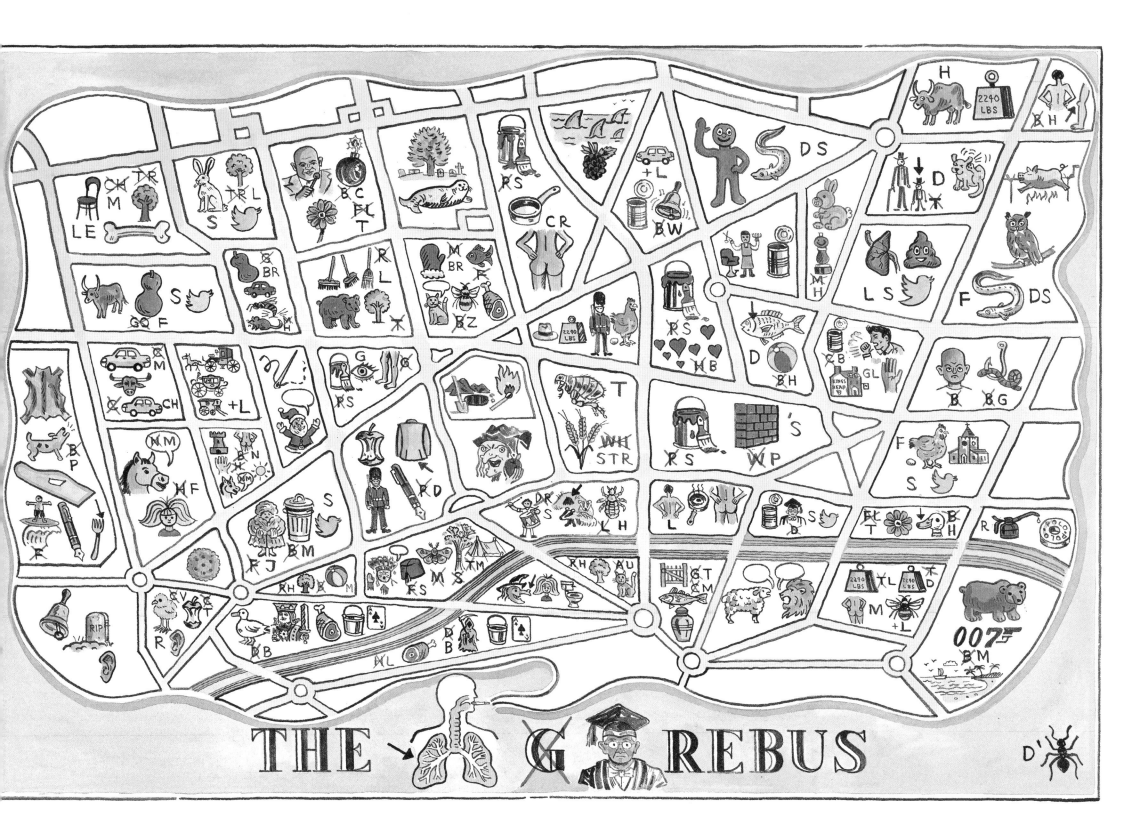

THE G REBUS

41

Like most Londoners, I moved here because I wasn't averse to the idea of being in the middle of a big crowd of people. Everything about the capital appeared to be the antithesis of the featureless, flat, windswept Fens where I'd grown up. I relished cultivating a proper Londoner's insouciance when pouncing through closing tube-train doors into a tightly crowded carriage of grumpy strap-hangers or when leaping onto the back of a moving No. 19 bus, hopeful of a cosy seat on the top deck, where I could smoke a fag on the way home in the traditional damp London fug.

Today, in my artist's studio slap-bang next to the usually teeming Liverpool Street station, I was just about to start colouring a busy cityscape of Leicester Square that I'd finished drawing in early March when it suddenly struck me that I was looking at a scene I might not see again for a very long time: a panorama from a pre-coronavirus age.

This week I put the finishing touches to my map of 'Viral London', a cartographic history of how the capital has been assaulted by various epidemics over time, from the plague of AD 664, which struck in monastic communities and seriously destabilised the church, to HIV/AIDS and the 'London Patient', the second person in the world to be cured of the disease.

The manner in which nearly 1,500 years of catastrophic epidemics have assaulted Londoners, transforming their behaviour and their urban landscape, is characterized not by the rolling out of major social projects but by observations and incidents that are anecdotal and poignant, like scenes from a developing story, the trajectory of which is always unknown and guaranteed to be grimly surprising. Walking home through a deserted City of London, H.F., the diarist of Daniel Defoe's *A Journal of the Plague Year*, witnesses a woman fling open her casement window to cry out into the night, 'Death, death, death!' Erasmus, visiting London during an outbreak of the 'sweating sickness', was too scared to drop by his lodgings for fear of the illness. And in 1918 revellers in town celebrated the armistice unaware that among the crowds the deadly Spanish flu was being spread.

Characteristic of my prints and drawings, this new depiction of Leicester Square, the apex of crowded London, contains all manner of embodied anecdotes, physical incidents and interaction, most of which now contravene the latest government advice on staying safe. The characters I'd imagined and inked are all hugging, fighting, nuzzling and standing a little too close to each other in the queue at the theatre ticket booth. Some are even shaking hands. *Quelle horreur.*

I started wondering how historic conflagrations and catastrophes, similar to our own Covid-19 pandemic, might have coloured the visual sensibilities of the artists of the past. For a start, we always imagine everything was so much more nasty back then. My artist heroes and predecessors all lived through and witnessed episodes of unforeseen viral assault. Bruegel's epic painting *The Triumph of Death* is underpinned by the ever-present risk of plague in the 16th century, Wenceslaus Hollar engraved views of London before and after the Great Fire that had helped wipe out the pestilence, and no Italian Renaissance artist working in Venice was safe from the 'French pox', aka 'Neapolitan bone-ache', aka syphilis. Former Leicester Square resident William Hogarth was fortunate to live in epidemic-free Georgian London, unless we include 'gin' on the list.

The paintings of all these artists are often teeming with people. La Serenissima of Guardi and Canaletto is practically a masked and wigged *Dové La Wally*, the beaky masks being filled with herbs to ward off the pestilence. If the peasants in Bruegel's *Kermesse* country-dance paintings didn't succumb to some serious gut-ache from the rather unhealthy counter surfaces on display, then the risk of viral transfer, as they dance about tightly packed and very drunk, would definitely have done the job of laying them low in the Low Countries. I'm left wondering if my depictions of packed public spaces will have to serve as a perpetual description of how we used to be, or whether I should start fashioning a more Watteau-esque kind of crowdscape, where figures waft elegantly past each other, at arm's length, with a courtly grace conferred by mandatory social distancing.

Leicester Square, Sloane Square, the Royal Exchange and all the other places I've drawn will never appear quite the same again after the cessation of the war on Covid-19, any more than might have been the case after the Blitz. Reconstruction of 'life as normal', expressed through genre painting and visual depictions of the everyday circus of the social whirl, may have to be styled to include manners of behaviour and social codes that at first might seem alien and absurd, but which when normalized over time will make the characters in the picture who are not playing the game really stick out from the crowd.

VIRAL LONDON

Hand-tinted print
56 × 76cm (22 × 30in)

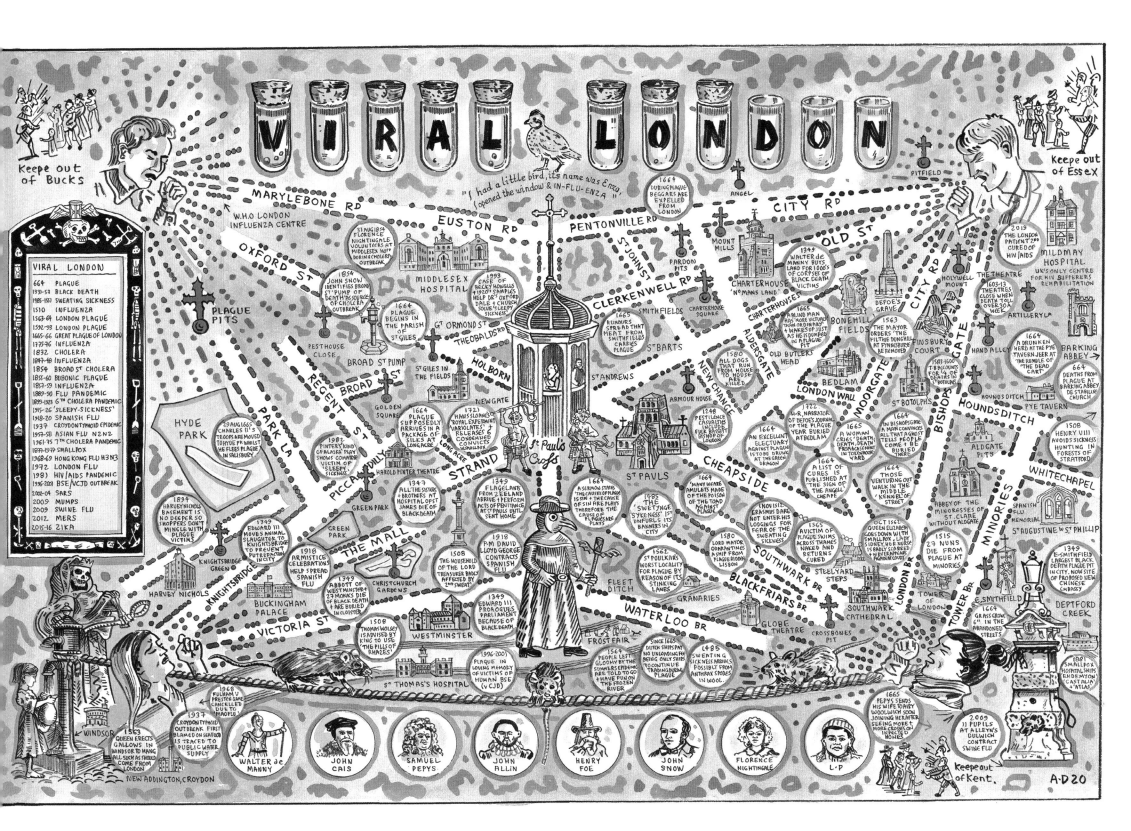

ICONOCLASTIC LONDON

Whether due to conquest, protest, dogma, lunacy, obsession, rightful anger, prurience or just because they were in the wrong place at the wrong time, London's statuary, monuments and works of art have throughout the ages found themselves the victims of all manner of deliberate, destructive actions.

Even this depiction of 'Iconoclastic London' appears to have been the target of a wanton 'map tearer', who in their ripping activities has succeeded in revealing the stories of London's various encounters with the Iconoclasts, just as acts of destruction are often claimed as a creative, liberating force deployed to reveal greater truths and expose stagnant or corrupted belief systems.

Sometimes we see our public facilities assaulted by well-organized 'political' campaigns, such as in the case of damage wrought by the Suffragette movement or the IRA, both of whom directly targeted Royal Mail postboxes, with paraffin-soaked rags in the case of the former and explosive devices with the latter. On other occasions our monuments are the victims of a type of collateral damage by dint of their proximity to popular points of public dissent and the sudden assumed guilt and culpability they possess as accessories to any crimes that happen to be in the process of being protested against at that moment.

Who knows what it was about sculptor Frank Dobson's *Woman and Fish* on Cambridge Heath Road, East London, that led to it becoming the target for continuous acts of vandalism? Or why the 1797 bust of a woman above the entrance to a former Huguenot soup kitchen was deemed fit for chipping off with a cold chisel? In 1769, when a Lascar sailor attacked the statue of Queen Anne that stands outside St Paul's Cathedral, he was simply whisked off to the madhouse.

Not caring about the edifices of one's immediate environs as they crumble, decay and turn to dust could be viewed as being as serious an act of civic irresponsibility as might a failure to smash the same monuments as part of a zealous and well-meaning campaign to redress the inequalities and wrongs wrought a decade, a century (a millennium?) ago.

It is possible that the status of statues as end-point embodiments of particular political projects, or as containers for the successful coalescence of how history happened to play out, are, in their immovability and reticent triumph of ideas, too much of a provocation in the face of activists whose inflexible beliefs and tenets are similarly (if metaphorically) set in stone.

The assuredness of the protesting reformer's belief eventually sanctioned by the law saw Bishop Nicholas Ridley having to be physically restrained, so keen was he to pull down the tomb of John of Gaunt. And sanctioned by nature; acid rain, in the case of the Rhodesian High Commission, aided the custodians of its hated sculpted figures of Jacob Epstein's *Ages of Man* frieze by finally eroding and emasculating them.

A story discovered just too late for inclusion on the map of 'Iconoclastic London' is that of the Bryant and May matchgirls, who in 1888 went on strike over poor pay and conditions and whose descendants every year, in an act of perpetual iconoclasm, paint red the hands of the statue of the then Prime Minister William Gladstone in Bow. As long he's there and they continue to do so, we'll know why.

ICONOCLASTIC LONDON

Hand-tinted print
56 × 76cm (22 × 30in)

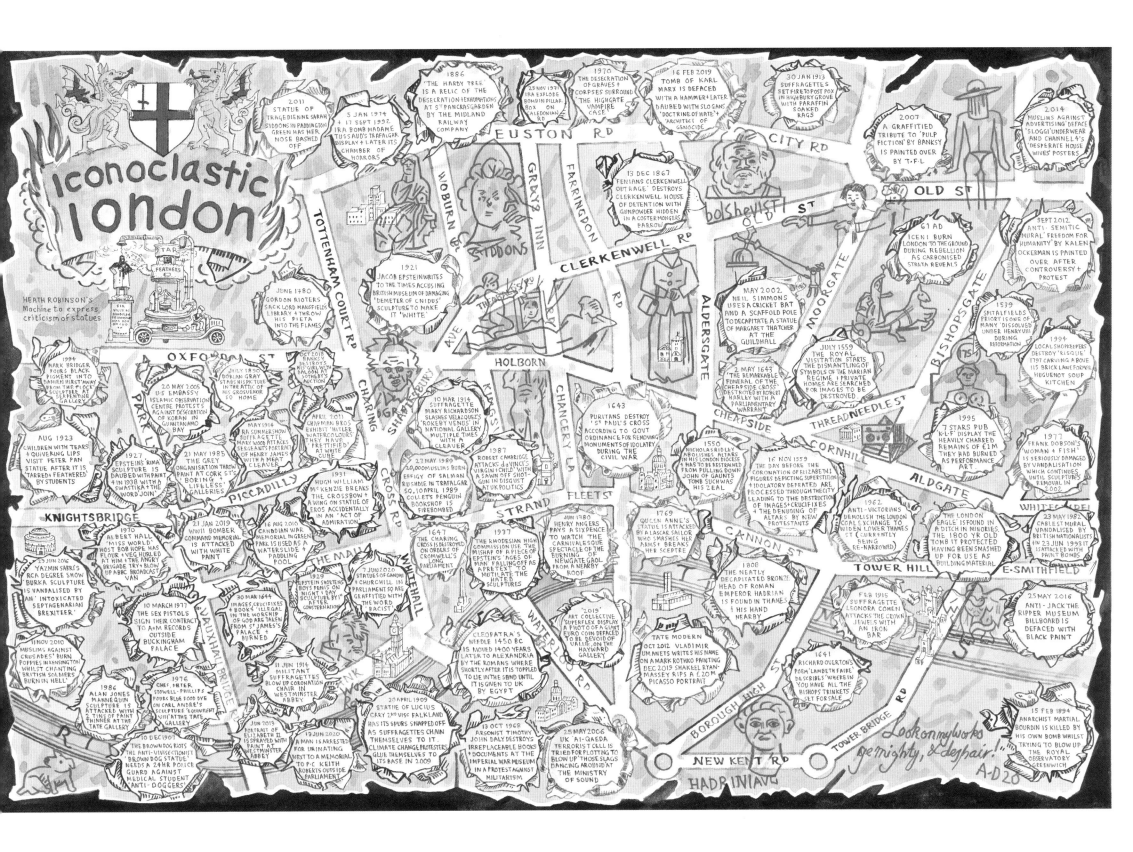

THE LONDON CALENDAR

'The London Calendar' map was created in anticipation of 'The Great Unlocking', when after a year and a half of officially imposed isolation and Covid restrictions, the population of the capital and most of the rest of the UK would emerge blinking into the sunlight like bears coming out of hibernation. Some displaying the additional ursine qualities of extra hairiness and big tums.

Having had such freedoms returned, however, did not necessarily mean a return to normal; the strangely arch phrase 'the new normal' came to denote, in Orwellian fashion, something that was not normal at all, in a similar mode to the famous French protest about the British system of monarchy: 'But does it work in theory?!' 'The London Calendar' of events was destined to hang on the kitchen wall as a mere advisory notice.

Never mind! All the events on the London calendar depicted across the map at the appropriate locations, listed according to when they occur, have been occasional victims to the odd plague, or been cancelled due to world wars and global conflagrations, but continue nonetheless because, being mostly folk celebrations, they have their roots in the strange and arcane popular history of London and the seasonal traditions of the country and the cycles of nature in general. A harvest festival is not going to be postponed to a more convenient time when all the turnips will have rotted in the ground and all the honey been stolen by the bears.

Thus, the London calendar lives in pulsating, corporeal fashion in the beating hearts of all the capital's citizenry who, without prompting, will feel in their bones that it's time to don the lozenged cloak of Merry Andrew, the fancy antlers of the Highgate ceremony of the horns, or will know instinctively that the time is nigh to head into Soho for the annual waiters' race, to Hackney for the clowns' church service or to the Tower of London for the annual washing of the lions on 1 April.

THE LONDON
CALENDAR

Hand-tinted print
56 × 76cm (22 × 30in)

THE LONDON CALENDAR

THE UNIVERSAL LONDON UNDERGROUND MAP

I t is often claimed that there are over 250 languages spoken in London. 'The Universal London Underground Map' dispenses with the need to produce 250 different travel maps by deploying, as it does, the common communication tool of the pictogram, or icon. Akin to the modern 'emoji', the message of the pictogram renders meaning in a clear and direct manner, appending specific signifiers to that which they signify within a context defined by a particular history and geography.

Thus travellers from all over the world furnished with a reasonably comprehensive knowledge of the place into which they are about to descend – of its peoples, its particular history and its popular and folk culture – will have no trouble at all when getting about.

Furthermore, their journeys on the London Underground network will be far more edifying when augmented by certain factoids and specific information, such as the part played in the building of Bayswater by Edward Orme (whose portrait denotes Bayswater tube station), the famous former presence of flamingos in the rooftop gardens (until, allegedly, drunken yobs threw them over the parapet) at High Street Kensington and the charming English folk tale by which Paddington not only refers to a major railway terminus but also to a Peruvian bear (*por favor cuida de este oso!*).

THE UNIVERSAL TUBE MAP

Hand-tinted print
56 × 76cm (22 × 30in)

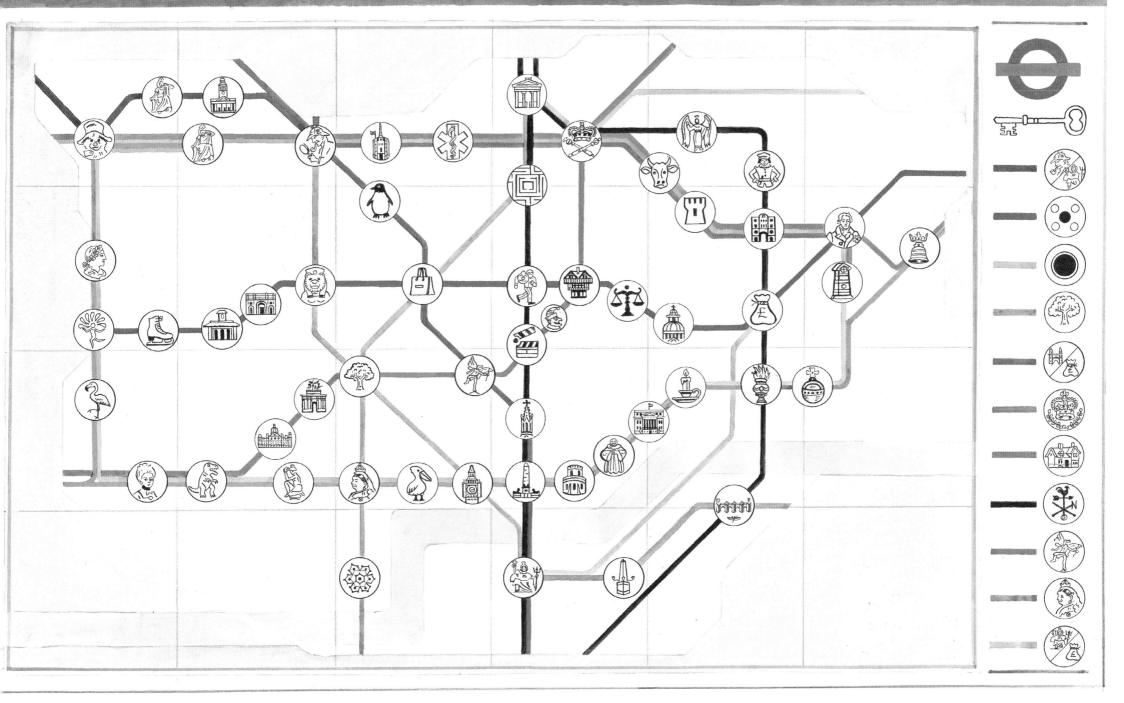

THE CRITIC'S PERPETUAL DRINKING CALENDAR

Some people never need an excuse to have a drink. For those less fortunate, 'The Critic's Perpetual Drinking Calendar' offers 365 of them (and a bonus tipple for a leap year).

Occasions for toasting, such as the Queen's birthday, the anniversary of Trafalgar or the invention of the drinking straw, are joined by opportunities to toast the lives of the famously bibulous on what would otherwise be just another day. Legendary devotees of the tap, bottle and jigger such as Dorothy Parker, Sultan Selim II and Arthur Rimbaud are each afforded their own special day, colour-coded according to a well-stocked cellar of varied libations.

'The Critic's Perpetual Drinking Calendar' makes a fine and fitting practical adornment for all the best barrooms, cocktail lounges, dining rooms, offices, kitchens, bedrooms and all other places of worship.

Cheers!

Don't mind if I do!

THE CRITIC'S
PERPETUAL DRINKING
CALENDAR

Lithograph
56 × 76cm (22 × 30in)

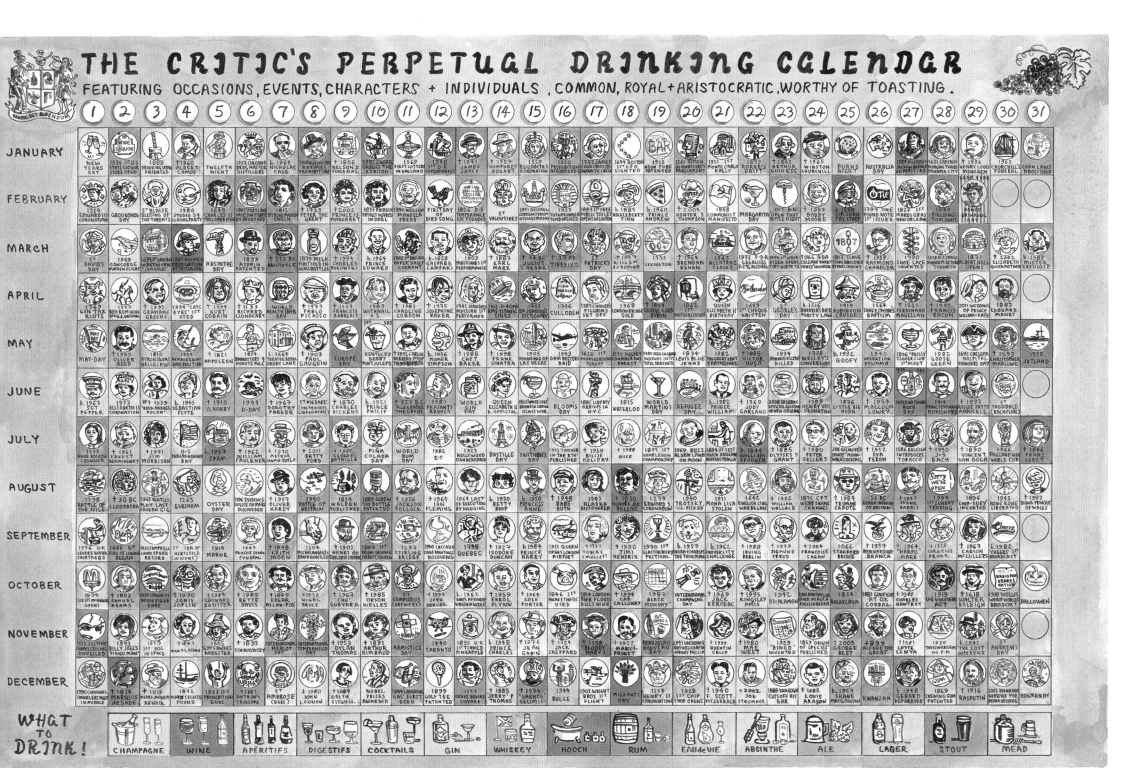

51

JEWISH SPITALFIELDS

Unlike the Spitalfields of the pre-Reformation church community, the French Protestant Huguenots and the Irish weavers, 'Jewish Spitalfields' is a historic incarnation of a particular London enclave that is well within living memory.

For many, London's East End will forever be synonymous with three things: the 'rag trade', beigels and the questionable manners of the perpetually harassed waiters at Bloom's famous kosher restaurant on Whitechapel High Street. Though fresh in the memory of Londoners who grew up there, and vivid in the anecdotes of their descendants, one can still feel the remaining traces of this old world slipping away into the mists of the past.

Encapsulating the western fringes of the 'Jewish East End' around Spitalfields Market and Brick Lane, the map of 'Jewish Spitalfields' takes as its model an original paper map place-setting from the now-defunct Bloom's kosher restaurant. Expanded and redrawn, the new map retains all the famous faces and places of the original Bloom's napkin map, but adds to it a new layer showing the shops and businesses, restaurants and general hangouts of the East End's Jewish community.

Sandys Row Synagogue.

BLOOM'S JEWISH SPITTLEFIELDS

Linocut and hand-tinted print
76 × 112cm (30 × 44in)

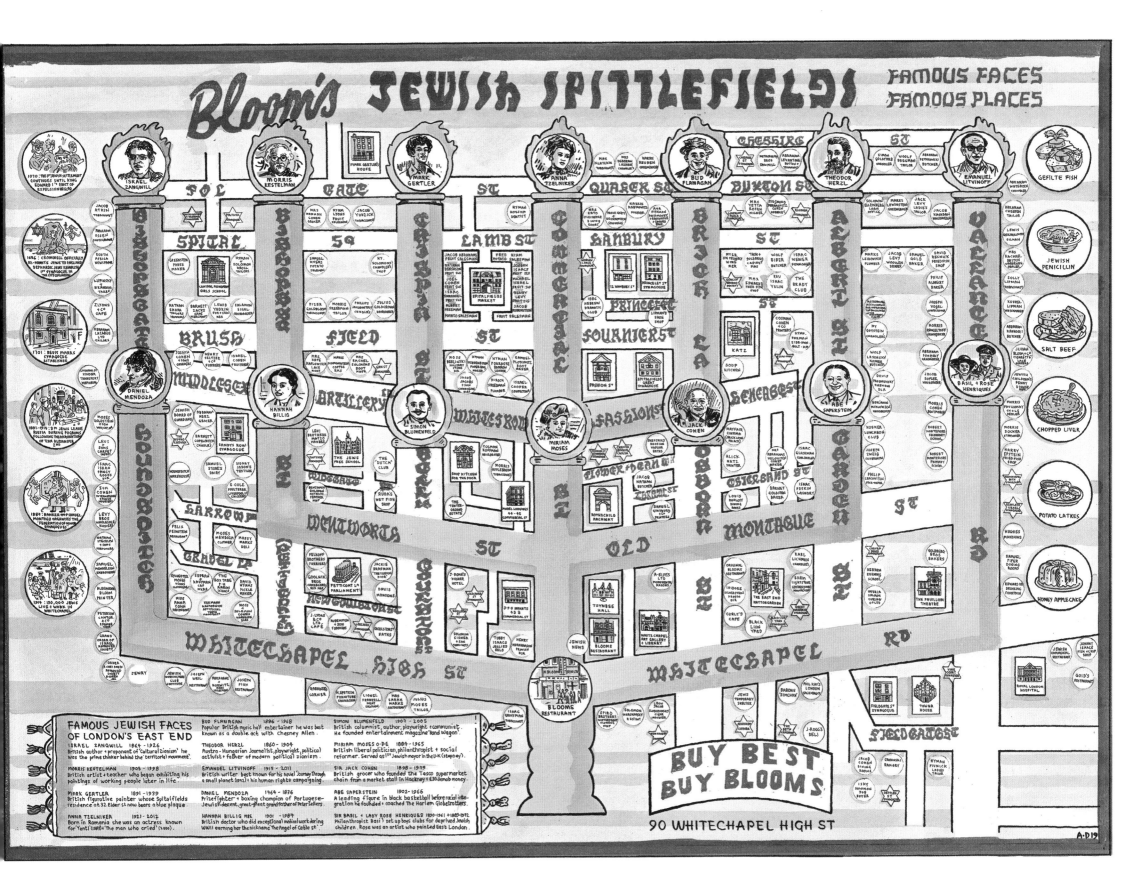

Bloom's JEWISH SPITTLEFIELDS — FAMOUS FACES FAMOUS PLACES

BUY BEST BUY BLOOMS

90 WHITECHAPEL HIGH ST

FAMOUS JEWISH FACES OF LONDON'S EAST END

ISRAEL ZANGWILL 1864 - 1926
British author + proponant of 'Cultural Zionism' he was the prime thinker behind the 'territorial' movement.

MORRIS KESTELMAN 1905 - 1998
British artist + teacher known for his paintings of working people later in life.

MARK GERTLER 1891 - 1939
British figurative painter whose Spitalfields residence at 32 Elder st now bears a blue plaque.

ANNA TZELNIKER 1921 - 2017
Born in Romania she was an actress known for 'Yentl' (1983) 'The man who cried' (2000).

BUD FLANAGAN 1896 - 1968
Popular British music hall entertainer he was best known as a double act with Chesney Allen.

THEODOR HERZL 1860 - 1904
Austro - Hungarian journalist, playwright, political activist + father of modern political zionism.

EMANUEL LITVINOFF 1915 - 2011
British writer best known for his novel 'Journey through a small planet' (1972) + his human rights campaigning.

DANIEL MENDOZA 1764 - 1836
Prizefighter + boxing champion of Portugese-Jewish descent, great-great grandfather of Peter Sellers.

HANNAH BILLIG MBE 1901 - 1987
British doctor who did exceptional medical work during WWII earning her the nickname The Angel of Cable st.

SIMON BLUMENFELD 1907 - 2005
British columnist, author, playwright + communist. He founded entertainment magazine 'Band Wagon'.

MIRIAM MOSES O·B·E 1884 - 1965
British liberal politician, philanthropist + social reformer. Served as 1st Jewish mayor in the U.K (Stepney).

SIR JACK COHEN 1898 - 1979
British grocer who founded the Tesco supermarket chain from a market stall in Hackney + £30 demob money.

ABE SAPERSTEIN 1902 - 1966
A leading figure in black basketball before racial integration he founded + coached The Harlem Globetrotters.

SIR BASIL & LADY ROSE HENRIQUES 1890-1961 + 1889-1972
Philanthropist Basil set up boys clubs for deprived Jewish children. Rose was an artist who painted East London.

A·D 19

53

CLUB ROW THROUGH THE AGES

My old east London studio has been a home, a library, a dining club, cocktail bar, birdcage, love nest, 'gallerette', city desk, publishing hub, a refuge from and for chaos, a dosshouse and an outpost for many foreign artists on manoeuvres … and quite a bit of art was created there too.

In May 2019 my landlady sends in the wrecking ball.

Artist's studios come in many shapes and quantities of cheap square footage. More often than not they wield the contingencies that stand between the artist and what they had hoped to achieve before adapting to the circumstances of their surroundings. The presence of the studio underpins a lot of great art just as the weather tells the sailor what to do with his boat. Lucian Freud turned a Paris hotel room into an etching studio using the sink as an acid bath. Andy Warhol's studio was a factory before, during and after his tenancy, and Picasso's *Demoiselles d'Avignon* look as if they could be crowded around the warmth from the last burning chair leg in the artist's freezing Bateau-Lavoir atelier. Given a 'dream studio', many artists never make a decent painting again.

When I left art school the standard off-the-peg, graduate-issue studio usually meant a whitewashed chipboard box in a warren of similar padlocked cells of creative genesis. Each of these turps-infused Petri dishes of the new art was occupied periodically by someone doing whatever they had to do while listening to Radio 4 or Joy Division on headphones between the McJob or teaching work that enabled the important stuff to happen and success to arrive.

I properly discovered that this kind of artistic apiary wasn't for me after helping transform a rundown East End hat factory into a studio complex with several other artists. All of us possessed the requisite drive, enthusiasm and surfeit of ideas to succeed in our artistic ventures, but were so painfully lacking in basic carpentry skills and unaware of how trifles such as local business rates worked that the process resembled an episode of the DIY kids' TV show *Here Come the Double Deckers*. Within minutes of the timber arriving for the construction of stud walls, built two inches narrower than the sheets of gyprock cladding as things turned out, one artist was already on their way to Whitechapel Hospital A&E having managed to smack herself hard on the forehead with a claw hammer.

The great camaraderie and bonding that resulted from everyone's common effort was valuable and felt good but at the same time it soon transpired that we had created a 'fishing village'. A self-contained community with all the attendant rivalries, gossip and vying for that necessity upon which every creative spirit most thrives … attention. On one occasion, having spent several weeks attempting to persuade

CLUB ROW THROUGH THE AGES

Ink and watercolour drawings
Each 42 × 29.5cm (16½ × 11¾in)

a smart, hip young gallerist to find the time to visit the studio to view my latest drawings, I was stunned on their departure to hear my artist neighbour engaged in a desperate and opportunistic attempt to sell their wares to the guy in the corridor.

I sought out artistic solitude, and the semi-derelict, vacant, mildewed, dank, gimped and squalid edifice that was to become the place from which I produced my artworks for the next 25 years was an all-time low.

But something about the place, its state of total grim decrepitude, told me that this wasn't the sort of building that would act as the setting for a sink-or-swim situation; if one's toes could already feel the bottom of such squalid waters, how could it?

The neighbourhood where I found myself had once been the beating oaken heart of the pre-First World War British furniture trade. The last vestiges of the veneer houses, cabinetmakers, brass fittings-mongers et al. who had made Shoreditch famous throughout the world were now planed down to a sliver, and in their place was a vacuum akin to the adjacent City of London's burst financial bubble. An odd assortment of artists, designers, musicians and a rubber-clothes 'dipper' now occupied an equally eccentric collection of ex-industrial spaces. In this grotty but convivial enclave there was always a ready surfeit of suitable space for the realization of whatever project one happened to have on the go.

It had always been my habit, whenever occupying a new place, to set as my first task the cleaning of the windows. Letting more light into a space is like giving it life and animation, and in the case of dodgy buildings in dodgy areas it also sends a signal to the locals that you're not going to be a dodgy one.

I painted a very handsome number 15 in a font of my own invention in black paint inside a grey circle on the wall of my new place, which, with the newly gleaming windows, swiftly resulted in approving comments from my new neighbours, relieved that the writing on the wall wasn't more graffiti of the 'Club Row = Disease' variety that had regularly marked out the area as the site of an unpopular dog market.

I spent several days a week as a gallery manager at a Mayfair Old Master picture dealers. One day at work, a couple of very incongruous Irish labourers stomped in and asked if anyone wanted to buy some big bits of wood. Before the prim, horrified Home Counties receptionist threw them out, I managed to secure the delivery of a stack of 19th-century 16-inch-wide oak floorboards ripped from the banking hall these builders were busy transforming into a handbag boutique.

I installed and made good the concrete-splattered floorboards upstairs in my new studio and, along with a hefty Belfast sink, a bath, a Calor Gas-fired kitchen, various salvaged features and the fancy parquet floor I had to lay as a muffling measure for a 'sound installation' during my studio's incarnation as 'The Gallerette', the place soon assumed a compact level of comfort akin to being in the captain's map room on the *Golden Hind*. When the head of watercolours at my

Mayfair gallery workplace spilled coffee all over his beautiful Japanese rough silk plan-chest cover I rescued it from his rubbish, made it new again and hung it across the alcove I'd created for my clothes.

During its time as The Gallerette my former studio mates from the former hat factory visited exhibition openings at what one of my colleagues in this venture suggested I should rename as the 'Laffably-Calder Gallery'. The idea of turning an old minicab office into a *Daily Telegraph*-listed art venue was preposterous but not without precedent. It was all a bit 'Dada'. Our first exhibition, 'We found a painting by Damien Hirst in a skip', managed to make the front page of *The Guardian*, completely enrage the eponymous artist and attract a potential lawsuit from the White Cube gallery, but it was not as much fun as the next show, a life-size photocopy of an elephant created by the Scottish artist Keith Farquhar.

For the opening party I'd attempted to secure the appearance of a real elephant on Club Row in the Victorian spirit that a visit from this animal brought with it a period of good luck for the neighbourhood. The circus told me that they'd be happy to come along with a young and willing pachyderm if I could shut the street and whack them some cash. It really wasn't as much money as it seemed to me at the time, or to the local council who I'd asked to foot the bill. Tower Hamlets Council's entertainments chap told me they could only stump up enough to pay for a camel. I left the council a message saying yes to the camel. They called back to tell me that, having reconsidered my plan, they might just about stretch to a donkey. Carlsberg Brewery were more forthcoming and sent over several crates of their special Elephant lager and a photographer for the Sunday morning launch party. I hadn't realized at the time that 'Elephant' referred to the high strength of the beer – 7.2 per cent! – and with a good number of bibulous Glaswegian guests it turned out to be a very good party.

One summer, friends attending the send-off for the gallery's artist in residence back to Berlin were asked to bring with them a bottle of whatever spirit was pencilled on their invitation. The artist had decided to stencil a pattern onto all the gallery walls with yellow spray paint; everyone involved was still trying to remove dried-up bits of enamel from up their now-hairless nostrils. At the party we drafted in the assistance of a cocktail waiter to mix the booze according to whoever came through the door next. The following morning the gallery was so full of forgotten coats, scarves, handbags and paraphernalia that it took most of the day to return all these items to their owners.

I soon realized that a studio could be a place whose doors could be as open to artistic endeavour as they were typically closed in the gestation of such.

I'd also decided at this time that publishing a daily newspaper would be a good idea. The result was *Donald Parsnips Daily Journal*, an A5, eight-page, 18th-century 'broadside'-style pamphlet that I wrote, illustrated and photocopied 100 times at Frank's 2p-a-copy shop. The irascible but supportive Frank was a pre-Bangla Town Brick Lane

boy with a concentration-camp tattoo and a host of nasty stories, unfortunately updated by that of a recent National Front firebomb through the back window of his shop.

I handed my *Daily Journal* to 100 unsuspecting Londoners en route from Bethnal Green to Old Bond Street every morning. It was to Fleet Street what The Gallerette was to Cork Street.

The newspaper garnered attention from the worlds of art and media and *Donald Parsnips* was soon featured on Radio 4 while appearing in galleries and museums around the world. During this time I installed a massive double trestle table in the studio that took up most of the downstairs and produced some big woodcut prints to sell as a way of paying off a mounting photocopying bill at Frank's.

As with most artist's studios, the primary activity in my place consisted of the mundane plodding labour that is strangely absent from most movies about artists. The reason for such editing out being that if cinemagoers were asked to watch what artists do most of the time they'd soon be fast asleep from boredom.

My thing was standing, day in day out, at a huge 8ft × 5ft drawing board with a brush and bottle of sepia ink and a selection of wooden boxes of various sizes on which I'd stand to reach particular bits of the picture. A friend of mine whose egg tempera paintings on gesso took months and months to complete told me that she'd spent 16 hours a day in the same spot for the past 15 years. She had cut out a square yard of linoleum on which to place her easel, painting stool and paint table. 'When I go,' she told me, 'they can hold up this lino at my funeral to show the congregation where I spent my life.' That's a seriously light carbon footprint.

Nobody died in my studio … in the car park over the road at the hands of a violent rent boy, maybe. Neither was anyone born there. But the ex-sweet shop, minicab office, chicken butcher, rabbit-house that I had turned into my 'atelier' was certainly the wellspring of numerous happy unions, including my own. The first time I saw my future wife she was sitting on the windowsill at the book launch of *Eh! A Novel*, a publication which aped Andy Warhol's typed transcript following 24 hours in the life of a single character around town with a tape recorder. In the case of the novel behind our meeting it was 24 hours in the somewhat dissolute life of an ex-art-school colleague whose reputation and behaviour had prompted his date that evening to ask the now Mrs Dant to attend the event as a chaperone.

Known as the 'artist' of the neighbourhood, I was often called upon to paint the shop signs for the local businesses. I'd done a rather nice series of planes, trains and trucks for the local Nigerian import-export company and a huge painting of a turtle on the doors of the next-door Livery Hall catering kitchens. The turtle is still visible as a shadow beneath the black doors of the trendy South Molton Street fashion house's eastern outpost who are my latest – and extremely sniffy – neighbours. One scene in *Eh! A Novel* sees the protagonist leaving the pub to go and collect my fee for a sign I'd done for a newly established

Bangladeshi fish wholesaler. He returned with my payment … in the form of a 20lb bag of basmati rice, which he'd subsequently had to haul around various pubs and clubs all evening.

Everything back then happened without smartphones, insta-messaging or Facebook. Social mediation meant the efficacy of the Royal Mail, *Time Out*, pub gossip and ESP.

Often things would happen according to some unknown, seemingly preordained logic. During a particularly lean period I remember the pay-as-you-go electricity meter had just ticked over and all the power had cut off. At that very moment there was a knock on the studio door. A film-crew runner in a high-vis vest stood outside with a £50 note in his hand. 'If I give you this, please can we run a cable from your sockets for our film unit?' he asked.

One of my neighbours who worked on history documentaries for the BBC made a film called *The Pamphleteer*, which recorded the thoughts of all the people who lived in my street linked together as readers of *Donald Parsnips Daily Journal*. Each of them were to have their own documentary; I was the first. It was odd viewing and gave form to the notion of how everyone's lives, their activities, their creativity, were completely interwoven with each other at the same time as being totally distinct.

This was the random and amorphous scenario made up of independent-thinking players that I'd hoped for, away from the partitioned multi-storey studio art-hubs.

It embodied the antithesis of what's currently called the 'echo-chamber'; we all seemed to be battling consensus. Every night at the local Owl and Pussycat pub everyone loved arguing with each other. They all seemed to have such very different ideas about what was right and what was wrong. It was funny.

I made a big drawing that depicted hundreds of scenes showing minor social transgressions. The unpredictable, awkward set-ups and pratfalls made famous by Bruegel, early cinema and possibly Eric Sykes. Today, everything – arguments, the artist's studio, the art world – seems so overwrought, so over-mediated and overly safe. Sealed with the planning committee's stamp of approval.

As with the total annihilation of many of London's more creaky and eccentric buildings, I wonder what's the best we can hope for from their complete disappearance. Though it always seemed rather nasty at the time, I still miss the 'edgy' conviviality of Charlie Wright's International Bar, now a hole in the ground, the 1940s ambience and technology of the Old Nichol letterpress factory, currently a very expensive fashion house, and sometimes I even miss the questionable hygiene and raggedy interior of Ron's Café on Redchurch Street, whose wafts of tallow are matched nowadays by the even worse pong of a stinky burger-chain outlet.

I'm currently in the middle of drawing a 'metaphysical' map of Cambridge. I was born there and I'm not going back. I just hope I can finish the picture before the bulldozers arrive.

GUARDIANS OF LONDON'S LOST RIVERS

If London's lost rivers ever had anything by way of protective gods akin to the famous Old Father Thames and his wife Isis then it's quite apparent in the loss of their fluvial charges that as river guardians they performed a very poor job indeed.

Or maybe not.

Maybe the role of such figures as personifications of animate powers is a task far more significant than that of mere 'figurehead', being that of 'fountainhead'.

London's lost rivers continue to flow, they still have identifiable sources and across this map of what is perhaps (if the torrent of guidebooks, novels, exhibitions, walking tours and maps is anything to go by) the city's worst-kept secret, they pour forth a healthy, 'gravel-rattling' flow.

By portraying the 'Guardians of London's Lost Rivers' in a more pragmatic and slightly prurient fashion and incorporating the histories of their meanderings into the personification of their sources, each is afforded an identity that needs no protector and no higher power other than that which keeps the springs gurgling. Even without such natural phenomena, Londoners themselves are ever keen to continue in the provision of ample streams of liquids.

The famously prudish Victorians, more so than the Londoners of any other era, must accept responsibility for a system that by efficiently hiding the passage of effluvia across the capital via its various natural Thames tributaries perpetuated the disappearance of these antique rivers. Many of the 'secret rivers' depicted on this map, most famously the Fleet and the Walbrook, had even by the Middle Ages become open channels of waste and subsequently culverted over. They were stinking receptacles of dung, dead dogs and all manner of industrial 'cack and mung'. The 19th-century engineer Joseph Bazalgette's network of vaulted sewers redirecting and putting to good purpose (if not in the navigable or angling sense) rivers such as the Tyburn and the Westbourne inadvertently highlights the foremost opportunity such historic natural features present, being that of carrying off that on which we do not wish to dwell. In such a context, would not any guardian of such an important passage personified be a bit 'pissed'.

So we have our cast of perpetually micturating masters and mistresses depicted according to the particular histories of London's lost rivers, incidents from which are detailed across this map.

The River Tyburn flows from the 'pissen-breeches' of the hanged man at Tyburn's 'triple-tree' gallows, London's antique place of execution; a gang of sailors provide the source of the Neckinger, close to London's docks; and one of the angels that the poet William Blake recounted appearing to him up in the branches of a tree in his garden looks after south London's lost River Peck.

GUARDIANS OF
LONDON'S LOST RIVERS

Hand-tinted print
56 × 76cm (22 × 30in)

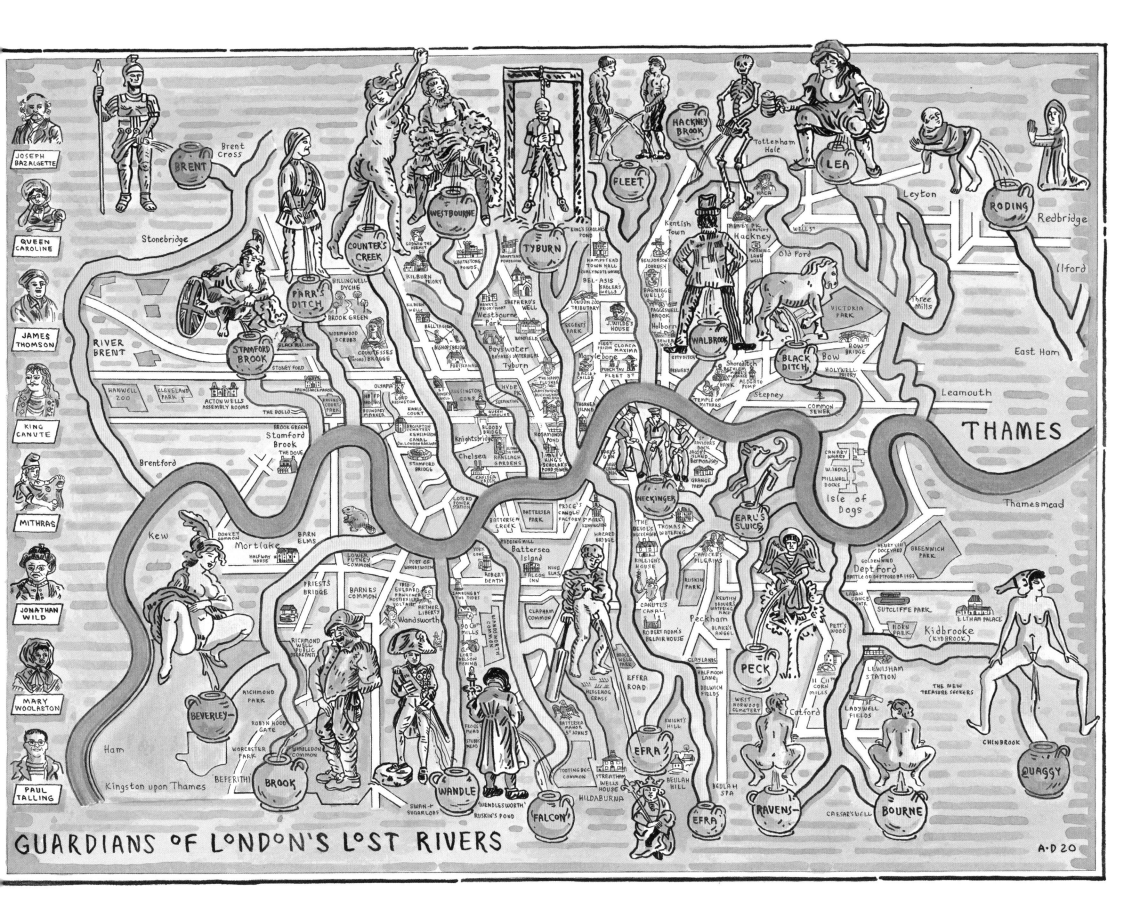

GUARDIANS OF LONDON'S LOST RIVERS

A·D 20

63

CIVIL WAR DEFENCES

The date of 23 June will forever mark a key moment in the history of a fractious and divided kingdom – long and protracted vacillation on the part of the powers that be and their favoured agents, a reticent and recalcitrant Parliament, and a threat to the status quo triggering any number of schisms: political, familial, religious and regional.

From outside, London was starting to look like a disputed central European town, as if the continent had claimed the island. The people of the capital, men, women and children, took to the streets voluntarily, marshalled and motivated by a fear of omnipotent and conclusive process. Wielding stakes and all manner of drums and ensigns, they set about digging in and advancing the cause of London. Calculated to divide and defend vested interests, London's lines of communication were erected as swiftly as they were removed.

As to the great responsibility handed over to Parliament in the hope that they would make the good and wholesome laws that the people of the nation expected, hopes were dashed. Instead of uniting a nation with righteousness and peace (which would have been a glorious thing to have done), what was provided was anarchy, corruption, division and dissatisfaction in what was from the beginning a provisional government, not truly representative of the people.

Enemies of the nation flourished under Parliament's protection. An immovable Parliament, as obnoxious as an immovable king, full of drunkards, tricksters, villains, whore-masters and godless self-seeking tricksters, no more capable of conducting the affairs of the nation than of running a brothel. Scum! And a truly elected scum at that. This is no Parliament.

The Conqueror pub in Shoreditch, which closed in 2006, sat fortuitously on the route of London's Civil War lines of communication.

'THE SINK OF ALL THE ILL HUMOUR IN THE KINGDOM'

Hand-tinted print
76 × 112cm (30 × 44in)

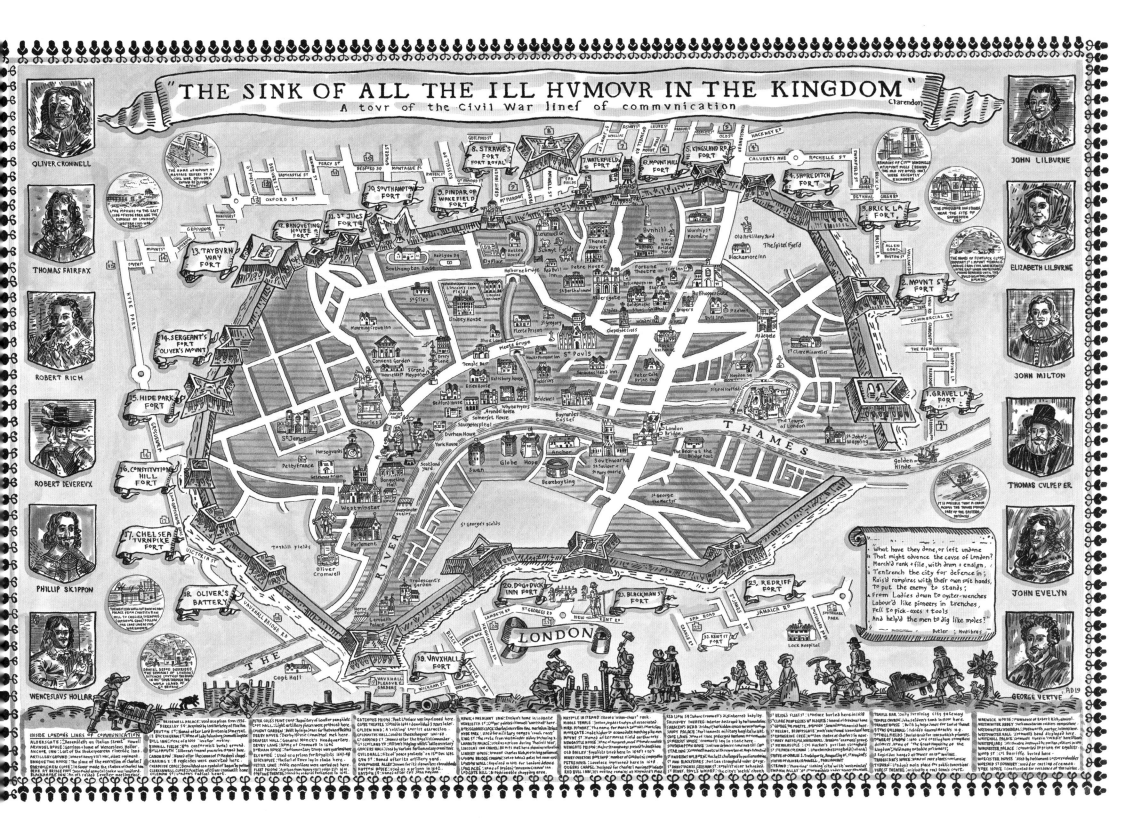

65

ARGONAUTICA LONDINENSI

To continue to assert in the modern world that the journey is as important as the destination is to foolishly avoid the fact that any point of departure, however boring or familiar, holds as much by way of adventure, surprise and arenas for incident and ordeal, from the start of the day until its end.

In addition, the most lengthy and predictably thrill-filled circumnavigation, the kind of voyage enjoyed vicariously by filmgoers, can be condensed into a similarly blockbuster length (two and a half hours, tops) period when such odysseys are relocated to the domestic terrain.

Taking its inspiration from Xavier de Maistre's 1794 fantasy *Voyage autour de ma chambre*, a parody of the grand travel narrative reset in the protagonist's room over a six-week confinement, and also obviously James Joyce's *Ulysses*, the 'Argonautica Londinensi' relocates the voyage of Jason and the Argonauts to central London.

Superimposing the street plan of the capital onto Abraham Ortelius's 1624 interpretation of the voyage of the *Argo* through the ancient world, the London *Argo* finds itself departing from Iolcus, wellspring of the map sellers of Cecil Court, through the clashing rocks at Aldwych and onward to Colchis (Watling Street at the heart of the City of London) and home of the Golden Fleece (public house).

In such a manner each episode in the journey of the Argonauts relocated finds the challenges set by each new location matching the incidents endured by Jason and company, so that London as the classical world finds:

- The rowing contest at Waterloo Bridge
- Hylas abducted by a water nymph at the Theatre Royal
- Thynias's deserted island at St Clement Danes
- The gates of Hades at St Paul's Cathedral
- Colchis at Bloomberg, Budge Row
- The River Ister at Oxford Street
- Circe's island at Marble Arch
- The *Argo* blown off course through Mayfair
- The wedding of Jason and Medea at St George's, Hanover Street
- The *Argo* landing at Libya/Lambeth and carried across the desert/ south London
- Talos the bronze giant at Westminster
- The water-carrying race at Trafalgar Square
- Santorini at St Martin-in-the-Fields

ARNOLD CIRCUS/OLYMPUS

Arnold Circus, near the artist's studio, is transformed into Olympus with the Greek Pantheon hidden in the trees. Three-colour linocut print created from flooring salvaged from one of the apartment blocks depicted.
1.5 x 1m (5 x 4ft).

ARGONAUTICA LONDINENSI

Hand-tinted print
56 × 76cm (22 × 30in)

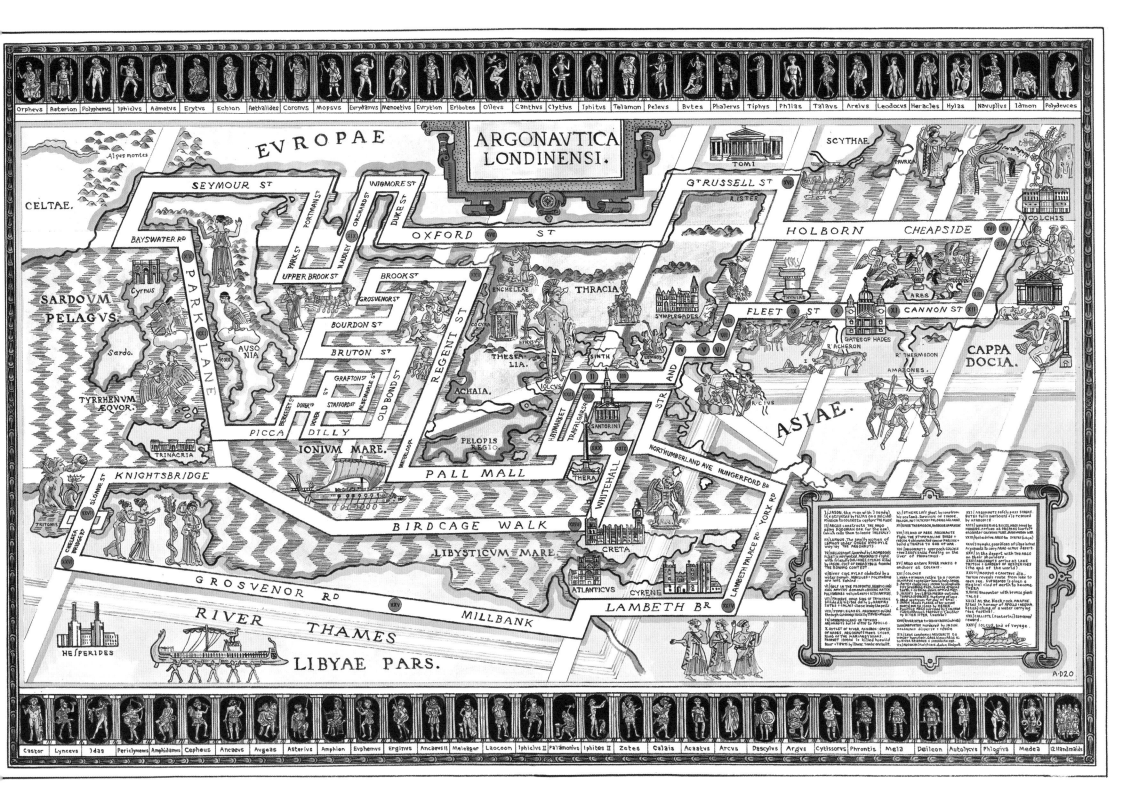

ARGONAVTICA LONDINENSI.

THE TRIUMPH OF DEBT

The global financial crisis of 2007–08 was documented minute by minute with vicarious, gleeful, forensic horror by the world's press, the language and phrases deployed taken straight from a narrative that seemed to have more in common with 'biblical cataclysm' than straightforward 'economic collapse'.

'The Triumph of Debt' was the title of an exhibition of drawings and prints that were made from within the thick of the 'catastrophe', while I was employed as the artist in residence for the high-finance publication *Spear's Magazine*. The magazine facilitated all manner of privileged access, info and gossip, just as it had done during a former assignment when I had been asked to undertake an artistic investigation into the then secretive world of the hedge fund manager. The exhibition 'The Art of Hedge' had proved an amusing and somewhat enlightening glimpse into an arcane realm that had for the most part been off-limits to the press and media.

For that particular assignment my editor had fixed me up with a number of tours of the Mayfair townhouses and sleek West London HQs of these privately run, largely unregulated funds of the new 'Masters of the Universe'. His theory was that – unlike a Dictaphone-wielding *FT* journo with 'snapper' in tow who stood no chance at all of gaining the trust of a bunch of slick stock-flippers who had no reason at all to speak with the press – an unassuming artist in a paint-flecked corduroy jacket with a sketchbook and a crayon could probably pass for days unnoticed in these new gilded chambers.

He was right, and this special access initially helped me to establish a visual archetype of the hedge fund manager's special world. 'The Art of Hedge' was a cutaway view of an office/lair/gentlemen's club within which the juggling of numerous telephone lines and trading screens was accompanied by oyster-shucking, specialist head-shrinking and particular background details glimpsed from within the domain of the 'Hedgies'. Zebra-skin rugs, supermodel receptionists, a fencing piste, mantelpieces laden with charity ball invitations and, in one office, a leather-topped desk, bare except for a large HM Treasury cheque book and fountain pen, were all detailed in my 'reportage' sketches from the field.

In a large architectural rendering of an Italian piazza, 'Palazzo Ponzi' depicted the exposure of the American financier Bernie Madoff's 'wealth management business' as nothing more than a $64.8 billion 'Ponzi scheme', the most audacious, flagrant and crooked financial swindle in history. Prominent individual victims of the Manhattan fraudster's dupe were depicted as busts in loggias on the façade of an edifice that, it turned out, was nothing more than that.

To assist with the construction of this drawing the magazine had linked me into a live, regularly updated feed that detailed each new investor/victim of Madoff's fraud as the investigation by federal prosecutors revealed the extent of the 'big lie'. From major international banks and Hollywood celebrities, the list was soon scrolling down to include firemen and primary school teachers in Liverpool.

The 2007–08 financial catastrophe, which was to be the subject of the prints and drawings of 'The Triumph of Debt', was a similar 'house of cards', 'perfect storm', or 'shit-storm' as the press described the events that were to climax in the bankruptcy of Lehman Brothers on 15 September 2008.

While sketching in Canary Wharf on the morning that employees of the fourth largest bank in the US were informed that they had, after all, not been 'too big to fail', I was struck by the relative calm of the scene, especially when set against the language of apocalyptic doom, scorching hellfire and total wipeout that was being wielded in the financial pages that day to describe the ensuing crisis. The most animated scenes to sketch involved the scrum of reporters, press photographers and cameramen that appeared from nowhere every time a Lehman employee exited the bank's revolving lobby doors. A few of them left carrying the mandatory brown cardboard boxes filled with the contents of desks and lockers, a pair of gym sneakers, signed American football or silver-framed family photo sticking up out of the top. One beleaguered but particularly photogenic employee, her blonde hair whipped up by the down-draughts of Canary Wharf's skyscrapers, appeared on the front of so many newspapers the following day shot from so many different angles that I was able to cut out all these press photos to create an animated view of her repeatedly spinning round and around a full 360 degrees.

When a few of the financial correspondents asked me what I was doing, it wasn't long before I was asked to appear on the type of news stations that artists rarely tune in to (the channels which usually had all sorts of numbers, acronyms and curt stocks-and-shares IDs scrolling back and forth, up and down and all over the screen in front of a perma-tanned, dentally enhanced transatlantic market analyst). After my interview on CNBC the presenter told me that he thought I was probably the first person to ever appear on the show wearing a herringbone tweed sports jacket and rust moleskin trousers.

The two big 'Triumph of Debt' drawings that formed the centrepiece of an exhibition of the same name depict the square outside Canary Wharf station, which had been the scene of Lehman Brothers' London meltdown, and Royal Exchange Square outside the Bank of England, which for generations has been the de facto spot to plant a news camera when covering stories concerning the world of banking and finance.

In essence the creation of both images as documents of this particular moment in financial history simply involved collecting all manner of over-exaggerations, dramatic phrases, hyperbolic comparisons and histrionic verbiage that journalists had deployed in their telling of the stories of the financial crisis and applying these in a literalist manner to the stage where the drama was taking place. The resulting scenes thus more resembled the kind of world seen in Bruegel's monumental painting *The Triumph of Death* as opposed to the rather flaccid failing realm of derivatives, de-leveraging, CDOs and CDSs and bailouts.

At the launch event for the exhibition I was slightly wary of how the battered and beleaguered bankers and financiers might react to being depicted as fodder for bear baiting or locked up in makeshift glass and steel corporate debtors' prisons. None of them seemed to mind in the slightest – after all, no one had been locked up or cast into a fiery pit. They didn't seem to have had to scale down on their sartorial outlay, they all still lived in the same houses, their children were all still ensconced in the same private schools, and they drank from a 'vodka luge' carved in the shape of an ATM from a big block of ice while joking about whose funds were carved into the depiction of a memorial 'to the fallen'.

Earlier in the year, in the middle of a 'wealth management' award ceremony held at Sotheby's auction house, which happened to be hosting a viewing of gilded and glossy Damien Hirst sculptures, news of US President Obama's refusal to 'write down' his nation's mortgage debts saw the galleries swiftly empty as mobile phones started ringing and pinging all over. The subsequent auction nonetheless made record sums, and moulded Perspex awards for 'Best Family Office' and 'Best Debt Re-packager' no doubt still grace the onyx mantelpieces or mega-basement 'man-cave' trophy cabinets of Chelsea and Holland Park.

It is probably safe to say that the types of punishment promised for kings and nobility depicted in Bruegel's *Triumph of Death* made little impact on a bunch of individuals who were unlikely to ever have their pomegranates snatched from them even if they did something really stupid. Similarly, the apocalyptic language used by the press in describing the financial crisis is unlikely to have spooked the current Masters of the Universe into considering buying a smaller car, having a 'staycation' or signing their kids up for the local comprehensive.

TRIUMPH OF DEBT, ROYAL EXCHANGE SQUARE

Ink drawing
102 × 152cm (40 × 60in)

National

Art

The Hogarth of hedge funds offers a glimpse into a hidden world

Artist spends six months documenting the mysterious lives of the wizards of finance

Charlotte Higgins
Arts correspondent

The men and women (or, let's face it, largely men) who are making fantastic fortunes working in hedge funds are often credited with tickling the art market up to its current giddy heights. So there's a pleasing neatness in the idea of an artist being commissioned to turn his scrutiny on to the "hedgies" themselves.

Adam Dant was commissioned, appropriately enough, by Spear's Wealth Management Survey (a quarterly magazine aimed at that special breed of humans known as high-net-worths) to document the professional lives of the mysterious creatures who, behind closed doors in Mayfair and St James's, engage in abstruse activities such as short-selling and leverage.

Dant, whose studio is on the fringes of the City of London in Shoreditch, spent six months in offices in the financial world – in particular taking a rare look inside the HQs of hedge funds, which have acquired a formidable reputation for secrecy.

He spent time, in particular, in the office of Clareville Capital, based in Westminster. Though his resulting works are an amalgam of a number of offices, his piece The Art of Hedge, showing a cutaway of a Georgian house on the fictitious Hedge Row, draws extensively on the sketches he made at Clareville, which is chaired by former Tory treasurer Lord Marland.

Dant, whose manner in the drawings is broadly Hogarthian, made visual inventories of the sort of items – and people – inside such offices. He also spent time following the hedge-fund managers around their chosen haunts – "though, alas, I never got invited on anyone's yacht".

He saw the hedgies disport themselves at Annabel's nightclub and private gambling establishments such as Crockford's in Curzon Street and the nearby Aspinall's, founded by John Aspinall, perhaps most famous as a chum of Lord Lucan. He saw them quaff cocktails at Harry's Bar on Mount Street, and buy up art at Sotheby's, Christie's and the best contemporary art galleries. "It's a very small world," said Dant. "They buy art from White Cube, because it's a reputable name. If you buy your shotgun from Purdey and have your shoes bought in Jermyn Street, you buy your art from White Cube."

The Art of Hedge shows the scene at a notional hedge fund office. Outside in the street is parked the standard-issue Mercedes, with its door being held open by the chauffeur, "who looks like maybe he's a Serbian," said Dant. Towards it hurries a hedgie, Blackberry attached to his ear. He is holding out a Christie's catalogue, whose cover image is a Damien Hirst. ("Every office must have its spot painting," says Dant, "and perhaps a Cy Twombly and Warhol.") His secretary is in hot pursuit, brandishing a diary, a White Cube invitation and an invitation to join the Olympics committee – an item he noted among the possessions of Clareville managing director David Yarrow.

A delivery man knocks on the door. He is holding an Aspinall's hamper – routinely sent round by the establishment the morning after a wealthy gambler has had a particularly bad night at the tables, according to Dant. The fictional firm is called Zeus – Dant noted the penchant among funds for borrowing nomenclature from classical mythology, such as Clareville's Pegasus fund. In the reception area to the right are "supermodel" administrators "laughing lamely at a hedge fund manager's joke". In the background are stacks of luxury-brand carrier bags and a crate of Chateau Latour. In the room at the lower left a stressed-out hedgie lies on a coach speaking to his personal psychiatrist – a detail inspired by an article about a hedge fund managers' shrink in Spear's. A hedgie consults his fencing instructor, and at the back, a cocktail waiter shucks oysters.

Upstairs, in the left-hand room, is a humidor with the best cigars and a baby-foot table. The walls are adorned with samurai swords and a shark's head. "It's always very aggressive, male stuff," said Dant. "And they really do regard the Art of War as their bible." He is referring to the 2,500-year-old Chinese manual on military strategy by Sun Tzu. At the back of the room is a wooden honours board "like one in a public school or Oxbridge college" showing the successes of the firm. "I saw one of these in a hedge fund office – it started in 2005 and had room to go up to 2028," said Dant.

Over the fire is a bust of Michael Douglas as Gordon Gekko, from the 1987 film Wall Street. In the next room the hedgies are at work, with phones draped at precarious angles about their faces "just like in Wall Street. I thought this was a cliche; in fact it is really what they do," said Dant. "Plus, they wear braces." In the corner a chap wearing tailored pyjamas takes a nap, waiting to do battle at some unearthly hour in the Asian markets.

"I wasn't doing this as a great sociological exercise," said Dant, "but I wanted to do drawings of things that really haven't been looked at before."

The Art of Hedge is at Robilant and Voena, Dover Street, London, from November 12-16

'Every office must have its spot painting, and perhaps a Twombly and a Warhol'

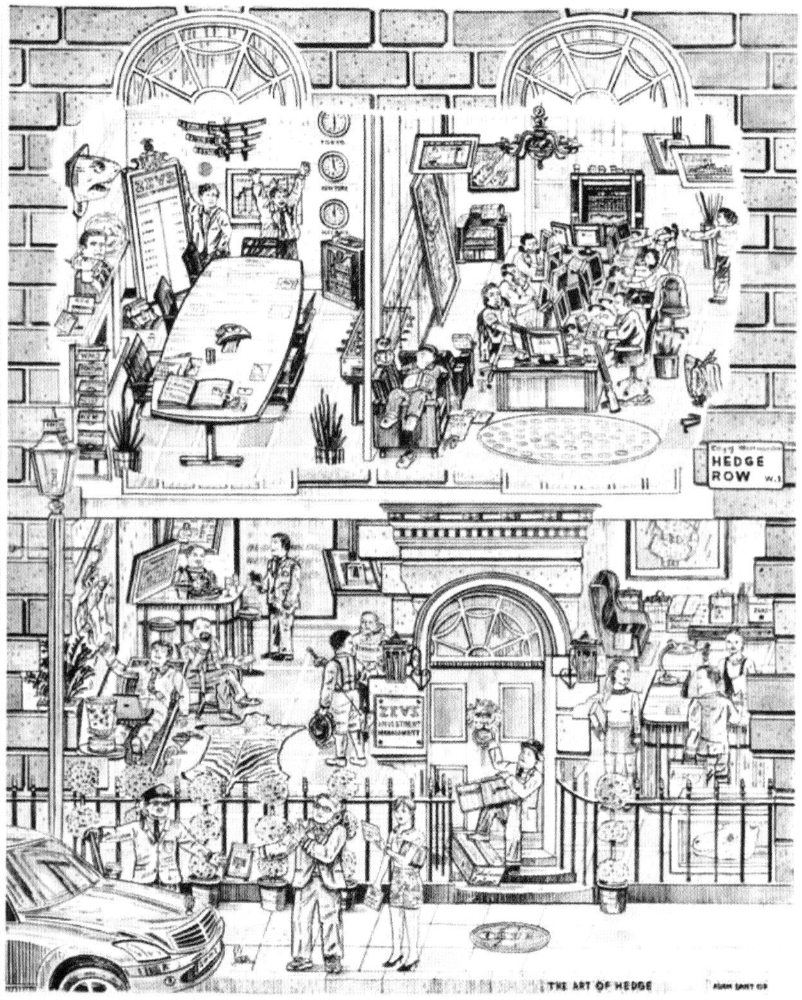

The house of Zeus
The Art of Hedge with its hedgies, secretary, supermodel administrators, personal psychiatrist, fencing instructor, cocktail waiter, and stacks of luxury-brand carrier bags and a crate of Chateau Latour. Left, a detail from Hell, where all bad fund managers go

---

ADAM DANT | THE ART OF HEDGE

This is what really Big Money looks like

ELIZABETH RENZETTI
erenzetti@globeandmail.com

Hedge Heaven: Dant observed in grand investment-fund offices to come up with his Hogarth-like drawings.

LONDON

On Monday evening, if you happen to be a member of the "new global class of the super-wealthy," you might find yourself upstairs at a posh art gallery in Mayfair, in order to have a good laugh at yourself.

The London gallery, Robilant and Voena, is old-school posh. This week, its halls will echo with booming laughter, the clink of champagne glasses and possibly even the rude trill of an unsilenced phone as several hundred bankers, lawyers, fund managers and their sleek consorts gather to peer at themselves in a series of new drawings by artist Adam Dant called *The Art of Hedge*.

Dant, a respected artist and winner of Britain's Jerwood Prize for drawing, spent much of the last year sitting in the corner of various grand offices in Berkeley Square, observing the largely secretive world of private investment houses and hedge-fund managers. The resulting drawings are a satirical peek into a hidden world where new money is married to ancient privilege: Traders make deals with hunting rifles by their sides, bankers pass out with laptops on their chests and martini glasses empty on the floor, shady chauffeurs wait to ferry their bosses to auction houses where they'll buy the latest hot art commodity.

"People are obsessed with what goes on in the hedge funds and in Mayfair, but no one really knows what does go on," says William Cash, editor of Spear's Wealth Management Survey, who commissioned Dant to produce the work. "He's a modest fellow, wears corduroy jackets," says Cash of the artist. "When a journalist walks in, none of these guys is going to talk. But if it's just some artist sitting in the corner, people don't get paranoid."

Dant's drawings aren't confined to "hedgies," that publicity-shy group of financial professionals whose work involves betting (not the kind found at the dog track) and the transfer of numbers that end in many zeroes. *Hedge Hell* groups together all manner of business people who have had a bad year – including Conrad Black.

If it seems a little cruel to kick a fellow when he's down, Cash protests, "I know Conrad quite well. ... He was a friend. But the reality is he's ended up in the rogues' gallery." Anyway, he says, the purpose is satire – "we're in Pope or Swift territory."

Spear's is a magazine, available only to those invited to subscribe, "exclusively for the members of the new global class of the super-wealthy."

Not a lot of lentil recipes in there, then. But it's a perfect publication for a city that's sick with money. For many reasons having to do with tax loopholes for foreigners, strong European and local currency, and doubts about U.S. markets, London is the hub for big swinging financiers. They appear in the papers when they run up £100,000 ($197,423) dinner tabs, when they order cases of champagne not to drink but to spray around nightclubs, and when they clean the shops of designer handbags at bonus time. The Frieze Art Fair, which draws the super-rich to London every October, is losing some of its sheen, mainly because artists grumble that you can't see the art for the bankers.

It is indeed a scene ripe for satire, but that's a bit tricky now that they're all in on the joke. Even better: They're creating the joke. The vulgar, money-obsessed climber used to be a figure of ridicule – think of Undine Spragg's parents in Edith Wharton's *The Custom of the Country*, or Augustus Melmotte in Anthony Trollope's *The Way We Live Now*. Now, there's a nostalgic haze around Tom Wolfe's *Bonfire of the Vanities* – people remember the warmth, not the fire.

I tried to get into see a performance of *Alex*, the new stage play based on the comic strip about a self-deluding City banker, but was turned away: The entire theatre had been bought up that night by a group of bankers and corporate headhunters as a treat for their clients. (This form of "corporate hospitality," which the play gently mocks, is huge business, and ensures that the business elite get good seats at every major event from rugby championships to the upcoming Led Zeppelin reunion.) I went to see *Alex* the next night, noting that I could have bought a "business-class seat" (with specially designed seat covers) for £40 ($79). I'm pretty sure I was the only person in the audience who worried about having cab fare home. It was a pinstriped crowd, and they howled at Robert Bathurst's energetic portrayal of the morally challenged Alex. Turn off your mobile phones and pagers, he told the audience members, before singling out someone in the first row: "You don't have a mobile? Loser!"

Alex lies, manipulates and weasels his way through his life, getting his secretary to schedule sex with his wife while he's orchestrating the downfall of his client, a troubled factory owner. (Bathurst is the only actor on stage; the clever conceit is that he interacts with other animated figures from the comic strip.) The satire hardly had a sting in its tail, though; audience members left chuckling.

Some of them may well be at the sale of Dant's drawings on Monday night. The series of prints called *The Art of Hedge* are selling for £10,000 ($19,700) each, and Cash doesn't expect there will be any left at the end of the evening. The subjects of the drawings – portrayed as drunk, sleeping, skiving off work – are expected to snap them up. "They're all very flattered," Cash says. "The only complaints we've heard are from people not included."

Cutting hedge Satirical art for City

It is a world of frantic phone calls, secretive meetings and spectacular rewards – and now it has its very own house satirist, **writes Peter Aspden**. A new exhibition of original art opens in London tomorrow, depicting the arcane – and occasionally ridiculous – world of the hedge-fund manager. The show, *The Art of Hedge*, by the British artist Adam Dant, makes fun of some of the beloved rituals of the "hedgies": sipping cocktails at Annabel's nightclub, buying contemporary art from auctions and shopping for luxury goods. In one scene, a fund manager speaks to his personal psychiatrist about his stressed life. Above the fire in the manager's office is a bust of their patron saint: Michael Douglas as Gordon Gekko. Mr Dant, whose Hogarth-influenced works are on sale at the Robilant and Voena gallery in Dover Street, spent six months in the offices of financial companies, quietly observing their habitat like a television naturalist. It is not only the hedge-fund types that take a hammering: bankers and lawyers are also depicted in their idiosyncratic surroundings. The images, commissioned by the quarterly magazine Spear's Wealth Management Survey, are proving attractive buys to the very people they satirise: the invitation to tonight's private view promised that the works would "add enduring value to your own private collection or your office".
The Art of Hedge, Robilant and Voena, November 13-16

Photograph: Daniel Jones

12/11/07 Financial Times

TRIUMPH OF DEBT, CANARY WHARF

Ink and watercolour drawing
102 × 152cm (40 × 60in)

MONEYSCRAPE

The 'Moneyscrape' bond share offer was launched from a folding card table outside the Royal Exchange in 1995.

Bearers of each of the 100 large woodcut print certificates would on the closing of the scheme after a year be entitled to a share of monies found in the street during the operation of the eccentric, peripatetic artists' newspaper *Donald Parsnips Daily Journal*.

The purpose of the Moneyscrape share offer was threefold. Firstly as a means of accruing funds to be put towards the *Daily Journal*'s mounting bills at Frank's 2p-a-copy photocopying shop on the Bethnal Green Road; secondly as a form of insurance policy for investors against inadvertent loss of coins; and thirdly as a work of fine art in its own right in terms of the Moneyscrape being a 'performance' and a 'product'.

The map of London according to the lost and found of small change gives a rough indication of the most profitable zones through which to stroll with an eye on harvesting coins from the pavement. It also serves to indicate the location of where particular coinage that became part of the Moneyscrape was found.

On the cessation of the Moneyscrape scheme, funds accrued throughout the year's flâneuring were passed on to investors at an annual shareholders' dinner and the operation wound up. Dividends were not distributed as a straightforward percentage of the total amount of found coinage but rather took the form of the actual found coins and the odd bank note being passed on to investors as tangible assets, their monetary value being ceded in favour of real value as physical works of art. The woodcut certificate, no longer serving its function as a promissory note, was also transformed into a work of art.

The history and presence of the Royal Exchange and in particular the activities of the early coffee houses had served as an inspiration for the birth and operation of *Donald Parsnips Daily Journal*. Just as early pamphleteers and the appearance of the first newspapers had their nascence in the mercantile and intellectual ferment of the 'Penny Universities', so *Parsnips Journal* sought to facilitate an operation along similar lines. News from stockjobbers, investors and insurers as well as discussions on the latest discoveries and general gossip were set down by the same 'pamphleteers' who passed these handbills to an eager readership operating with no mediation or middlemen, much in the same fashion as when London's streets were taken as literally being lined with gold.

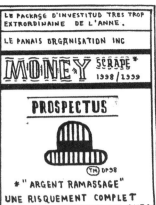

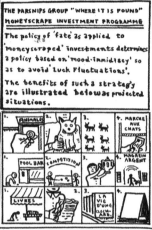

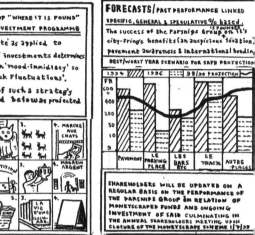

The Moneyscrape Prospectus (above). A map of the daily routes taken by the artist from East to West documenting particular 'moneyscrape premiums' (opposite, above left). The Moneyscrape Bond (opposite, below left).

The French Offer of Moneyscrape Bonds at Galerie Brighi, Passage des Panoramas, next to the Paris Bourse (opposite, above right). The London Offer of Moneyscrape Bonds outside the Royal Exchange (woodcut print, opposite, below right).

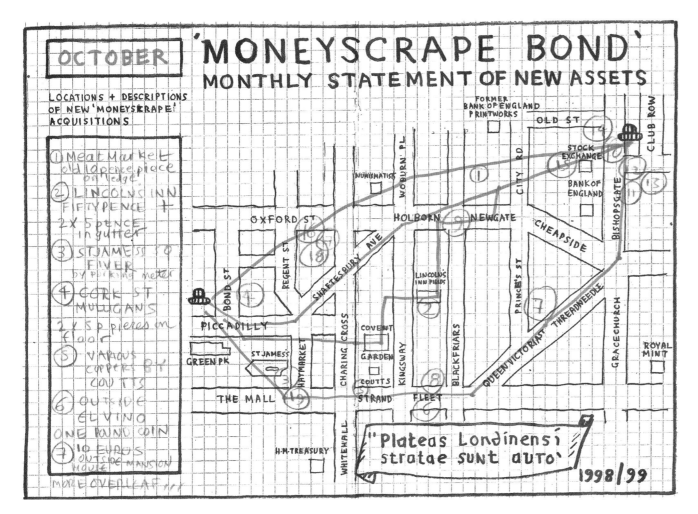

'MONEYSCRAPE BOND'
MONTHLY STATEMENT OF NEW ASSETS

OCTOBER

LOCATIONS + DESCRIPTIONS OF NEW 'MONEYSCRAPE' ACQUISITIONS

1. Meat Market old 10 pence piece on ledge
2. LINCOLNS INN FIFTYPENCE + 2 X 5 pence in gutter
3. ST JAMES'S SQ FIVER by parking meter
4. CORK ST MULIGANS 2 X 5p pieces on floor
5. VARIOUS COPPER BY COUTTS
6. OUTSIDE EL VINO ONE POUND COIN
7. 10 EUROS OUTSIDE MANSION HOUSE

MORE OVERLEAF...

"Plateas Londinensi stratae sunt auro"

1998/99

THE ISSUING OF THE MONEYSCRAPE BOND PROSPECTUS
PROVISION OF SECURITY AGAINST ACCIDENTAL LOSS & EVASION
EXCHANGE P¹LONDON FOR THE YRS 1998/99

BANCO PARSNIPS — DIEZ
MONEYSCRAPE ENDORSEMENT
10
DIEZ SNIPS
INTEGRITY HONESTY CONTINUITY OTHER
DONALD PARSINIPS

PRIME MINISTERS OF THE UNITED KINGDOM

In 1986, on the 300th anniversary of Number 10 Downing Street, the five living former Prime Ministers assembled after dinner for a formal photograph. Someone asked, 'I wonder, what is the collective noun for a group of Prime Ministers?', to which Harold Macmillan swiftly replied, 'A lack of principals'.

That those who have held such high office should cast such grave and jocular aspersions as to their own scruples does give serious legitimacy to the public's seeming general dislike of politicians.

Though newspaper lists of Britain's most popular and least popular Prime Ministers claim to put aside malleable and partisan arguments in favour of assessments of their effectiveness in office, one only has to check the masthead to discover why the top principal on one list finds themselves at number 53 on another.

While the popularity or unpopularity of some Prime Ministers polarizes opinions for obvious party-political reasons, others may have been unpopular in their day but are now relatively forgotten, such as Lord Goderich, who was described by George IV as 'a damned, snivelling, blubbering, blockhead'. History continues to be unkind to others. Lord North tops many lists as the most unpopular Prime Minister, both historical and current, the American War of Independence being an event that, like the Suez Crisis, the second Iraq War, the creation of the NHS and the other 'events dear boy, events' described concurrently on this timeline of UK Prime Ministers, eclipse the everyday travails that might mark out other periods of office as unremarkable.

As a popularity contest or even a beauty pageant, a fondness for wigs – or even Whigs – might swing opinion behind earlier incumbents, while a penchant for the more unpredictable and chequered biography might favour PMs of the post-war era. It's entirely up to the political opinions of the viewer whether they perceive things as getting better or worse when travelling this timeline, which provides a blank roundel at the end in which to validate said opinions.

PRIME MINISTERS OF
THE UNITED KINGDOM

Hand-tinted print
56 × 76cm (22 × 30in)

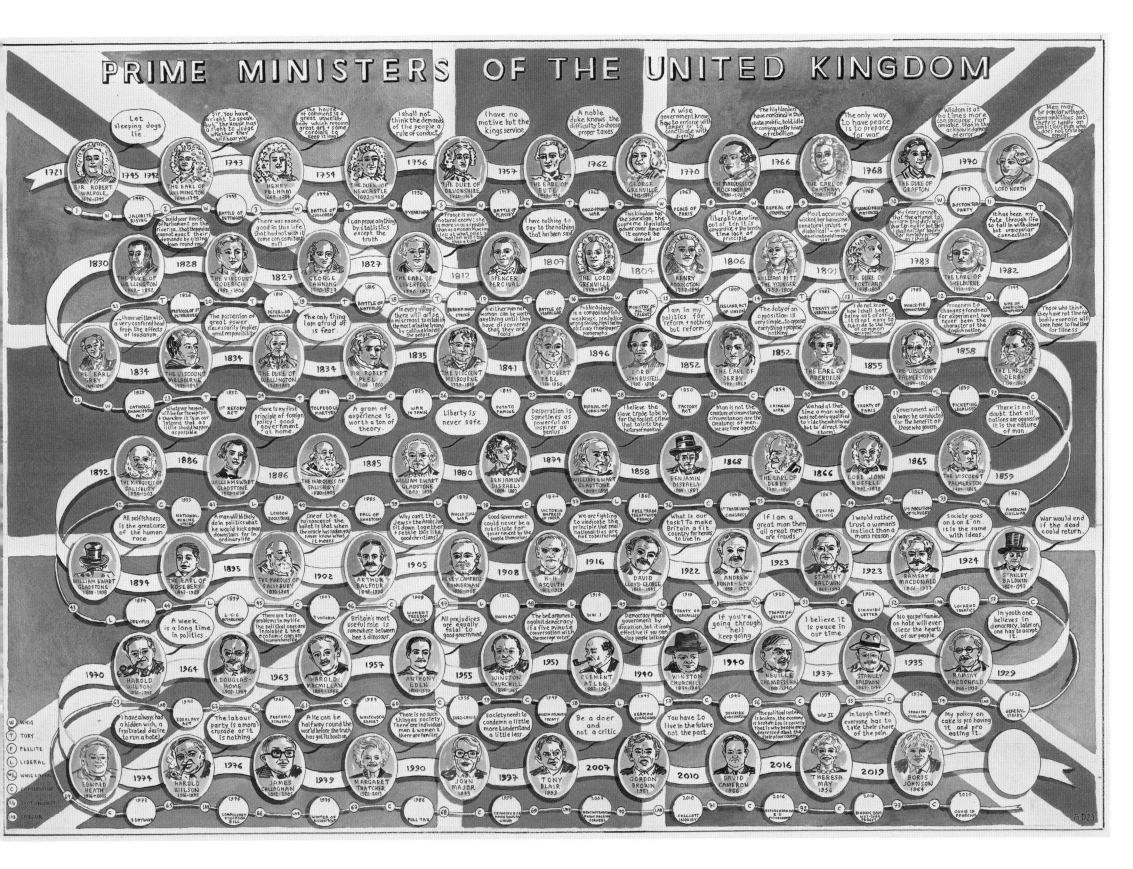

Presidents of the United States of America

Fairly straightforward, though dense with information, the timeline of 'Presidents of the United States of America' took as its inspiration a wipe-clean educational placemat from the collection of my daughter, then a toddler, which also included a political map of world, a chronology of the Kings and Queens of England, a French lexicon, an alphabet of animals and – rather ambitiously – a periodic table of the elements. Being half-American and a regular visitor to the thrift stores of various flyblown Texan 'piss stop' towns, such items took her fancy from among an array of curios that I as a Brit had only ever seen in movies and TV shows: familiar but alien objects such as college yearbooks, autoharps, letterman jackets, Davy Crockett hats, bear paws, stuffed rattlesnakes and guns.

As a child I too had had something of an obsession with the kind of comprehensive charts and timelines that my daughter was to cover in creamed spinach and rusk crumbs, though I collected them as posters rather than as wipe-clean adornments for the dining table.

I was not alone as a child in having an obsessive habit for collecting all sorts of things, possibly a consequence of being shut up in the cluttered box rooms of elderly relatives who still hadn't got round to disposing of the precious possessions of their own now-adult children. These dark, dusty attic rooms were filled with Edwardian schoolbooks, wooden educational toys, sets of Arthur Mee encyclopedias, scrapbooks of newspaper cuttings from the Great War, and – to the modern eye – very questionable histories from the farthest reaches of the British Empire.

My childhood bedroom was similarly crammed full of stuff. Apparatus for breeding butterflies, microscopes, specialist tools, fancy drawing equipment, electrical components, bits and bobs found while digging, and all manner of trunks and cigar boxes containing pinned insect specimens, pressed flora, labelled matchboxes full of sand from different beaches, animal skulls and the usual standard collections of stamps, cigarette cards and football stickers. Every breakfast cereal box contained something to collect. I seem to recall one brand coming with the promise of a real seashell lurking, unbagged and completely, tooth-crackingly hidden among the puffed and sugared product. One schoolfriend had a set of shelves at home to display his impressive collection of bottled air from various parts of the world, while another, possibly inspired by the Catholic schoolchild's familiarity with holy relics, had a small tin labelled 'spare parts' in which he kept teeth, scabs and (hopefully) little else, all (also hopefully) his own.

Though the timeline of the Presidents of the United States is not so thoroughly detailed as to inform the viewer about the childhood, education or peccadilloes of the first 44 residents of the White House up to and including Barack Obama, it does provide a map of all the US states, a chronology of historical events, the names of the vice presidents and their party political affiliations, as well a relevant quotation above their portraits. As the image was compiled prior to the next election year, a veil has been discreetly drawn over the 45th President of the United States.

Presidents of the United States of America

Hand-tinted print
76 × 112cm (30 × 44in)

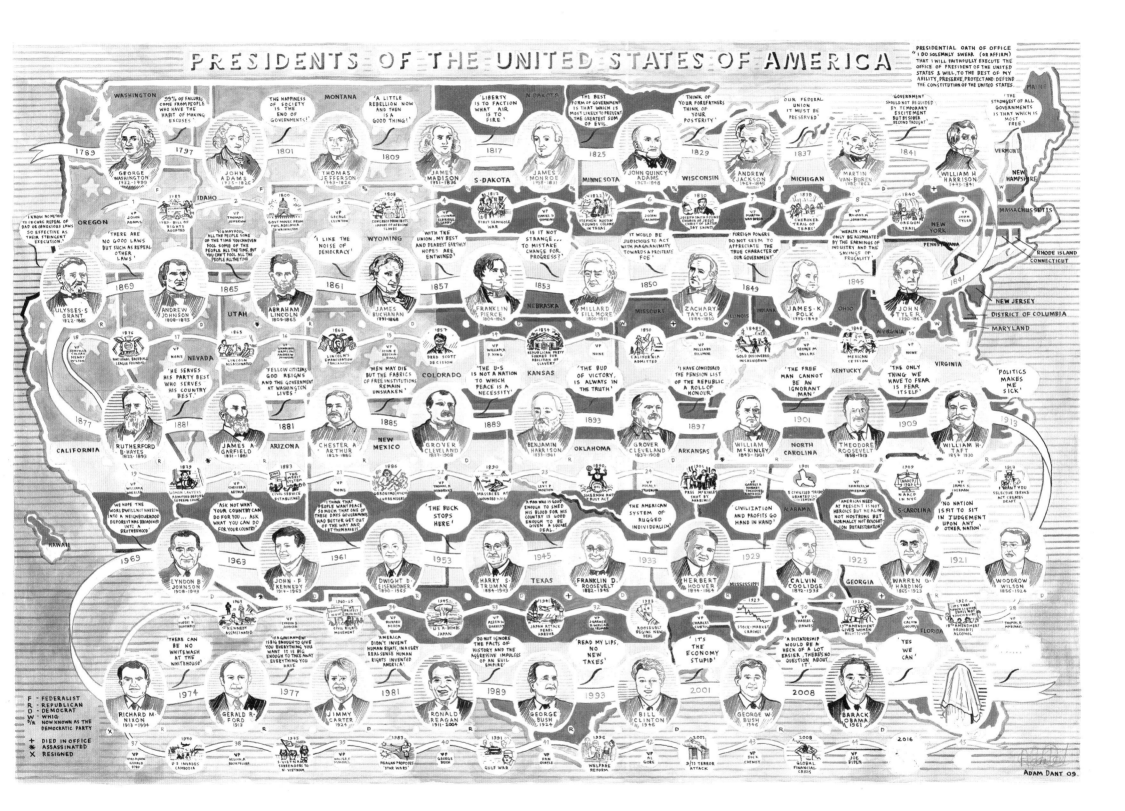

COCKNEY RHYMING AMERICA

The map of 'Cockney Rhyming America' renames all the states of the USA according to the bizarre and arcane argot of the east London costermonger, late of the parish of St Mary-le-Bow, Cheapside.

In the restrictive lingo of a community of questionable legitimacy, even according to its own tenets, the places of another distinct realm are given a new (and completely irrelevant) nomenclatural set, which is in essence of no use to either party.

The many examples of slightly embarrassing British variations on the American theme depicted below this map embody a continuity of transatlantic accord whereby our partners in the 'special relationship' really have no need of, do not care about and probably never even get to hear any opinion that anyone in the 'olde land' earnestly and expertly feels fit to broadcast about their 'Yankee cousins' (usually on *Channel 4 News*).

Aside from the great British success stories of Beatlemania, swinging London and the reselling of various other bits of American-born phenomena back across the Atlantic, most attempts by those in the peasouper-cloaked climes of Blighty to latch on to and emulate their far cooler ex-colonial cousins are best left out for the trash man. Transatlantic imposters found their apogee in the visible-Velcro costumes of the dire 1970s kids TV show *Animal Kwackers* (thankfully not depicted on this map), a bunch of *Banana Splits* wannabes who even purloined their name from the Barnum and Bailey-inspired biscuits (cookies).

COCKNEY RHYMING AMERICA

Hand-tinted print
56 × 76cm (22 × 30in)

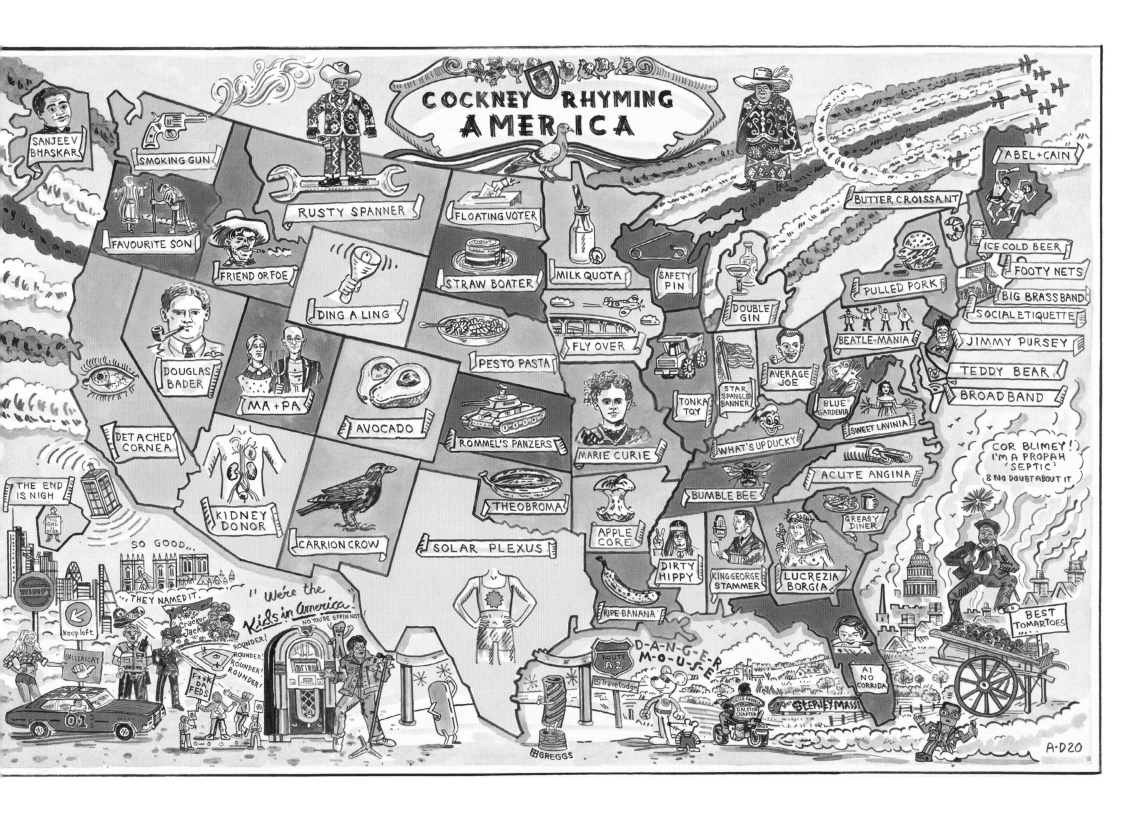

NEW YORK TAWK

N ew York Tawk is a visualization of New York City through a century of its slang, some recent, some with deep roots, all mainstream urban. There is no suggestion that this is comprehensive: New York has been one of slang's great centres since the mid-19th century and that kind of coverage would be foolish even to imagine. But I have taken some of the city's themes – social, professional, commercial, entertainment – and checked out some of the terminology. Enjoy.

PREPPY

Scoots money
Bruno Brown U.
Pencil-geek one who works more devotedly than their peers think fit
Reel in the biscuit to achieve a woman's seduction
Big Red Cornell U.
Get Chinese to succumb heavily to a drug
Gome a devotedly hard worker
Wonk an excessively hard worker; an expert
Mo a moustache
El zippo nothing
Cereb an intellectual
Weenie a hard worker
Fruit loop the small loop on the upper back of many shirts

THEATRICAL

Turtles 'resting' (i.e. out-of-work) actors
Oke fodder a hit show
Fannie a fan dancer
Terpery a dancehall
Stub-holders the audience
Pitman a musician
Milkman one who overacts

Hip flippers a 'hoochie-coochie' dancer
In the test tube a tryout for a new show
Cleffer a songwriter

YIDDISH

Schmeckel a fool
Futz a fool
Farchardet confused
Zetz a punch, often figurative
Schmutz dirt (misspelt on the map)
Nudnik a pest, a fool
Tummler the 'life and soul of the party'
Plotz to lose emotional control
Yentzer a nag
Doppess a layabout
A-k an old fool, from *alte-kacker* (old shitter)

IRISH

Tad an Irishman (from Thaddeus)
B'hoy a Bowery rough of c. 1850
G'hal the b'hoy's girlfriend
Pat an Irishman (from Patrick)
Rileyed drunk

Donnybrook a fight
Brannigan an argument
Pistareen a criminal who proclaims their innocence
Running for sweeny escaping
Gazoony an amusing person

BEATNIK

Roach last part of a joint
Dexies Dexedrine
Snazzy smart, fashionable
Cruddy unappealing, unattractive
Spodiodi mixture of cheap port and generic bar whiskey
Benny Benzedrine
Weirdsville a strange situation
Zoot suit a suit worn in the 1940s and 1950s, characterized by a long, draped jacket with padded shoulders and high-waisted, tapering trousers
Gasser something excellent
Schizo crazy
Goofball barbiturate
Hep-cat a sophisticated individual
Gone intoxicated
Steeazick a cannabis cigarette
M morphine/marijuana

GAY

Kiki a gay man or woman happy in both active and passive sex roles
Quiche attractive
Top in a sado-masochistic relationship, the dominant partner
Breeder a heterosexual
Read to assess a person's character/personality
Throw shade to malign, to humiliate
Spill the tea to retail a juicy piece of gossip
Glamazon a beautiful woman
Cakes the buttocks

COPS

Hocksheet a list of 'wanted' individuals
Creeper a burglar who robs houses when the owners are at home
Short eyes a child molester
Pussy posse the prostitute squad
Blue suit a uniformed policeman
Hot prowl a burglary while the property owners are present

Mopery + dopery a trivial offence
Bunco squad squad devoted to combating confidence tricksters
Wear the bag wear a police uniform
Perp a villain, a perpetrator

JIVE

Astorperious upper-class
Stuff cuff the padded cuff of a draped suit
Cackbroad boastful woman
Doo wah diddy an out-of-the-way place
Dusty butt a low-grade, unattractive prostitute
Eight rock a very dark-skinned person
Cracking but facking conveying hard factual information in the guise of jokes and humour
Granny grunt a mythical figure to whom otherwise unanswerable questions are referred
Pinktoe a white person
Miss Big Stockings an attractive, well-built, conspicuous young woman
Bama a peasant

Drape shape zoot suit
Kitchen mechanic a servant

LATINO

Tumbe a swindle, a confidence trick
007 a large folding knife with a wooden handle
PR Puerto Rico/Puerto Rican
Bollo-loco a promiscuous female, literally 'crazy chick'
Perico cocaine
Moreno a black person
Coño a fool, a general insult
Paro extremely intoxicated
Vato a gang member
Cheeba marijuana
Puto a gay male

MADISON AVENUE

Blipvert a very brief television advertisement, lasting one second
Pizazz style, glamour, ostentation
Run it up the flagpole and see who salutes to test something out by assessing responses
Adveka the advertising equivalent of Eureka! the moment of discovery

Skidmark a TV ad that stays onscreen during the entire programme

Leveraging integration team-based problem-solving

Sporno homoerotically charged images of sportsmen which tip the hat to a gay porn aesthetic

Pharmaganda medical product advertising that in its perceived bias is seen as propaganda

Programmatic the use of automation in buying and selling of media

HIPPY

Trippy disorientating, mimicking the effects of LSD

Heavy multi-purpose description, both positive and negative

Turn on, tune in, drop out slogan extolling psychedelic intoxication, associated with LSD pioneer Timothy Leary

Hash hashish

Grok to fully appreciate, coined by Robert Heinlein

Crash pad a temporary place to sleep

Trucking persisting

Groovy excellent, first-rate

Outasite first-rate, excellent

Dropout one who has also turned on and tuned in, a member of the counter-culture

Grass marijuana

ITALIA

Sit down a meeting (i.e. of crime bosses)

Paisan a fellow Italian

Chooch a fool, usually large, thuggish

Tenderoni an attractive young girl

Bodini the penis

Guido the stereotyped, ostentatiously macho, young Italian-American male

Goola the anus

Jabroni a failure

Capeesh an emphatic interrogative, 'you understand?'

Labonza the posterior

Tutti-frutti a male homosexual

Capo a criminal boss

Walyo a young man

BROOKLYN HIPSTER

Bronson a beer

Bust a moby to dance

Chipper a woman who's easy

Kale money

Tassel a girl

Frado an ugly guy who thinks he's good-looking

Jerry a stoner or hippie

Cronkite boy

Midtown uncultured or unhip

Deck a key word for most hipsters, similar in meaning to the antiquated fresh

Fin the opposite of deck, similar to outdated terms like 'wack' and 'lame'

Piece a cell phone

Juicer a ladies' man

MONEY

Potato possibly linked to such vegetable terms as kale, cabbage, lettuce, etc.

Shekels biblical imagery

Chump change a derisory amount of money

Cheese a possible play on another money term, 'bread'

Dooteroomus perhaps from 'duty' or the book of Deuteronomy

Piastre originally a small-denomination coin in various currencies

Long green the colour of dollar bills

Shin plasters originally low-denomination notes used on the frontier

Ducats an old name used in Europe

Paper notes

Spondulics from Greek *spondulikos*, a shell, used as early 'money'

Benjamin a $100 bill, from its picture of Benjamin Franklin

New York Tawk

Hand-tinted print
112 × 76cm (44 × 30in)

PARIS ARGOT

The particular slang of the 'City of Light' is here depicted in the form of a giant snail whose shell contains the outward spiralling arrondissements of Paris and all manner of generally filthy 'argot' concomitant with the proclivities of each.

Unlike the 'vulgar tongue' of its English equivalent, this particular *langue au gouttière* is no such thing, afforded as it is the status of being worthy of respectful study by L'Academie Française.

Thus, tourists using 'L'escARGOT Parisienne' will be assured a friendly reception from their 'Parisite' interlocutors.

Parisian abbatoir workers in
La Vilette, 1914

L'escARGOT
PARISIENNE

Hand-tinted print
56 × 76cm (22 × 30in)

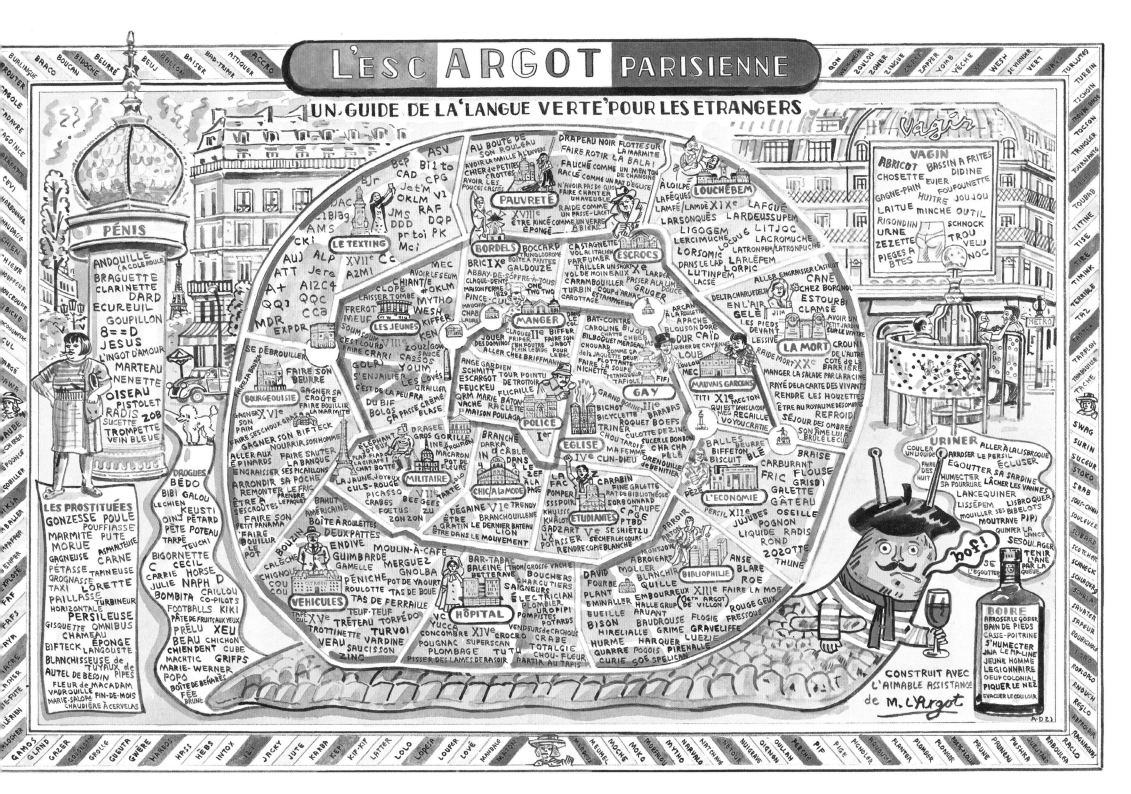

Using the Westgate of Oxford as a central axis point, 'The Oxford Meridian' project proposed the creation of a new meridian that incorporated the history of the new Westgate Shopping Centre site, the important place of Oxford in the world and a vision of Oxford as a city of the future.

The process of creating and establishing this new meridian involved the participation of the people of Oxford, whose unique contributions through exercises in 'speculative geography' – as well as via specific knowledge as to places on each of the eight possible routes of a new meridian according to the eight points on the compass – assisted in the creation of a type of 'visual timeline' display around the 360 degrees of the Westgate Shopping Centre archaeological site's hoardings.

Activities such as the making of three-dimensional globes (both celestial and terrestrial), an examination of the Oxford skyline (during night and day) from various high vantage points, the reconfiguring of lines of latitude, and the drawing of various stories in the form of 'timelines' generated a wealth of material from which to create a convincing model for a new 'cultural meridian' for the city.

An initial pamphlet that introduced the project to the citizens of Oxford was produced in a style that recalled and made reference to Oxford's unique history and its suitability for such an endeavour. Points of reference included the former scriptorium at Greyfriars, the English Civil War pamphleteering tradition, the geographical specificity of Oxford and its academic tradition.

The culmination of this artistic project resulted in the publication of a new 'Atlas' for distribution at the new Westgate Shopping Centre as well as other places worldwide that sit on the new Oxford Meridian.

In the course of mapping the strange and unusual connections that various random points on each of the eight possible meridians had with the city of Oxford, I was surprised to discover that the house in which I'd been born in the village of Swavesey outside of Cambridge sat directly on the line of the Greenwich meridian.

THE WESTGATE MERIDIAN

Printed pamphlet.

APPROACHING THE WESTGATE OF THE FUTURE   THE CREATION OF A NEW 'CULTURAL MERIDIAN'   THE PROPOSITION DISPLAYED AT WESTGATE

1. FROM THE WESTGATE OF OXFORD WE SET OUT TO EXPLORE THE CITY & BEYOND 'AS THE CROW FLIES', OUR INTENTION BEING THE CREATION OF A STORY FOR 'THE CITY OF THE FUTURE'

2. BY COMBINING THE ESTABLISHED CONVENTIONS OF 'THE TIME-LINE' (A FAMILIAR 'HOARDING' DISPLAY DEVICE) AND 'LINES OF LATITUDE' A NEW 'CULTURAL MERIDIAN' CAN BE REALISED WITH 0° SITED AT WESTGATE.

5. PRACTICAL ACTIVITIES THAT WILL TAKE PLACE WHILST WE ATTEMPT TO CREATE, LOCATE OR INVENT 'THE WESTGATE MERIDIAN' WILL INVOLVE THE LOCAL POPULATION AND WILL CONSIST OF...

a. ...A STUDY OF THE PANORAMIC OXFORD SKYLINE FROM VARIOUS ELEVATED VANTAGE POINTS AS THEY STAND ON LONGITUDINAL LINES EMANATING FROM A NOTIONAL QUADRANT AT WESTGATE, OVER THE HORIZON & BEYOND.

6. BEING THE 0° AXIS POINT FOR INVESTIGATIONS INTO A 'WESTGATE MERIDIAN', THE SITE OF THE NEW WESTGATE SHOPPING CENTRE'S SURROUNDING HOARDINGS PROVIDE THE 360° REQUIRED TO DISPLAY THE 'NEW GEOGRAPHY' OF THIS PROJECT IN A SUITABLE FASHION & CONTEXT.

7. IMAGES & INFORMATION SUCH AS TIMELINES, PANORAMAS, HISTORICAL NARRATIVES, VISIONS OF OXFORD DAY & NIGHT, CELESTIAL & TERRESTRIAL GLOBES ETC ON DISPLAY IN 360° MIGHT ALSO FORM THE BODY OF A PRINTED 'WESTGATE ATLAS' PUBLICATION.

3. THIS NOVEL CIRCUMSCRIBING OF THE GLOBE FROM THE POINT OF VIEW OF THE WESTGATE, OXFORD IS INTENDED TO ACT AS A COUNTER TO A PREVAILING HOST OF 'MEANDERING-WALKING TOURS' AND 'STANDARD HISTORICAL CHRONOLOGIES'

4. THE MANIFESTATIONS OF MANIA FOR OXFORD'S HISTORY, THE FRIARY, ROYAL COURT, ECCLESIASTICAL & ACADEMIC MATTERS, FANTASTICAL FICTION ETC, ALL CONTRIBUTE TO VIEWING THE IMPOSITION OF A 'CULTURAL MERIDIAN' AS A NATURAL & UNSURPRISING DEVELOPMENT.

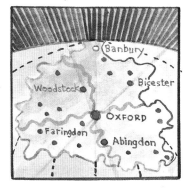

b. THE DECORATION OF 3 DIMENSIONAL GLOBES (CELESTIAL & TERRESTRIAL) WHEREBY THE WESTGATE SHOPPING CENTRE CAN BE POSITIONED IN RELATION TO OTHER SIGNIFICANT & RELEVANT CULTURAL AND HISTORICAL MARKERS.

c. THE CREATION OF VISUAL 'HISTORICAL TIME-LINES' USING INFORMATION DISCOVERED WHILST FOLLOWING NEW LINES OF LATITUDE AS THEY EMANATE FROM WESTGATE, RUN THROUGH OXFORD AND CONTINUE ON TO CIRCUMSCRIBE THE GLOBE.

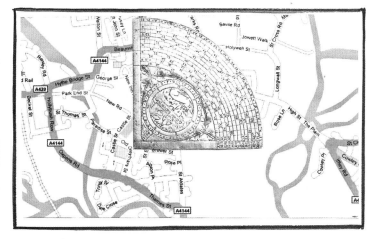

THE WESTGATE MERIDIAN,
NORTH–SOUTH

Ink and watercolour drawing
97cm (38in) diameter

THE WESTGATE MERIDIAN,
SOUTH-SOUTH-WEST–
NORTH-NORTH-EAST

Ink and watercolour drawing
97cm (38in) diameter

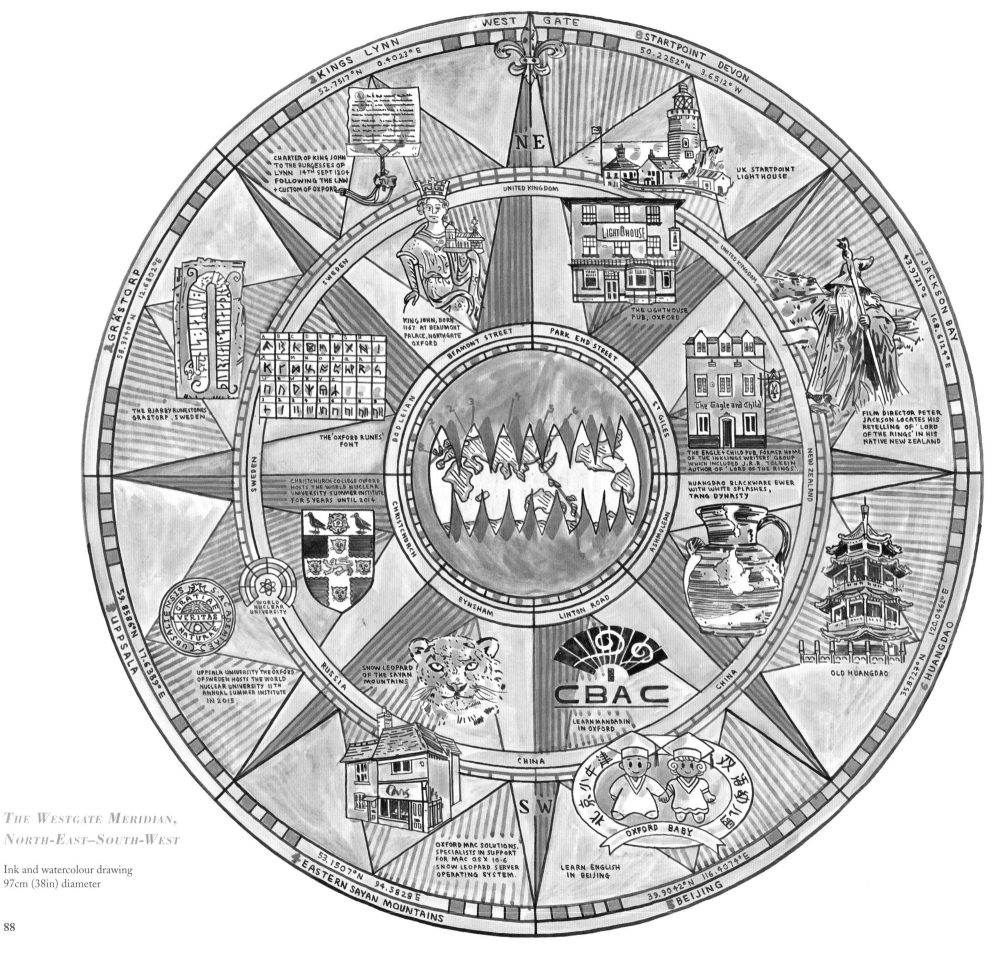

THE WESTGATE MERIDIAN,
NORTH-EAST–SOUTH-WEST

Ink and watercolour drawing
97cm (38in) diameter

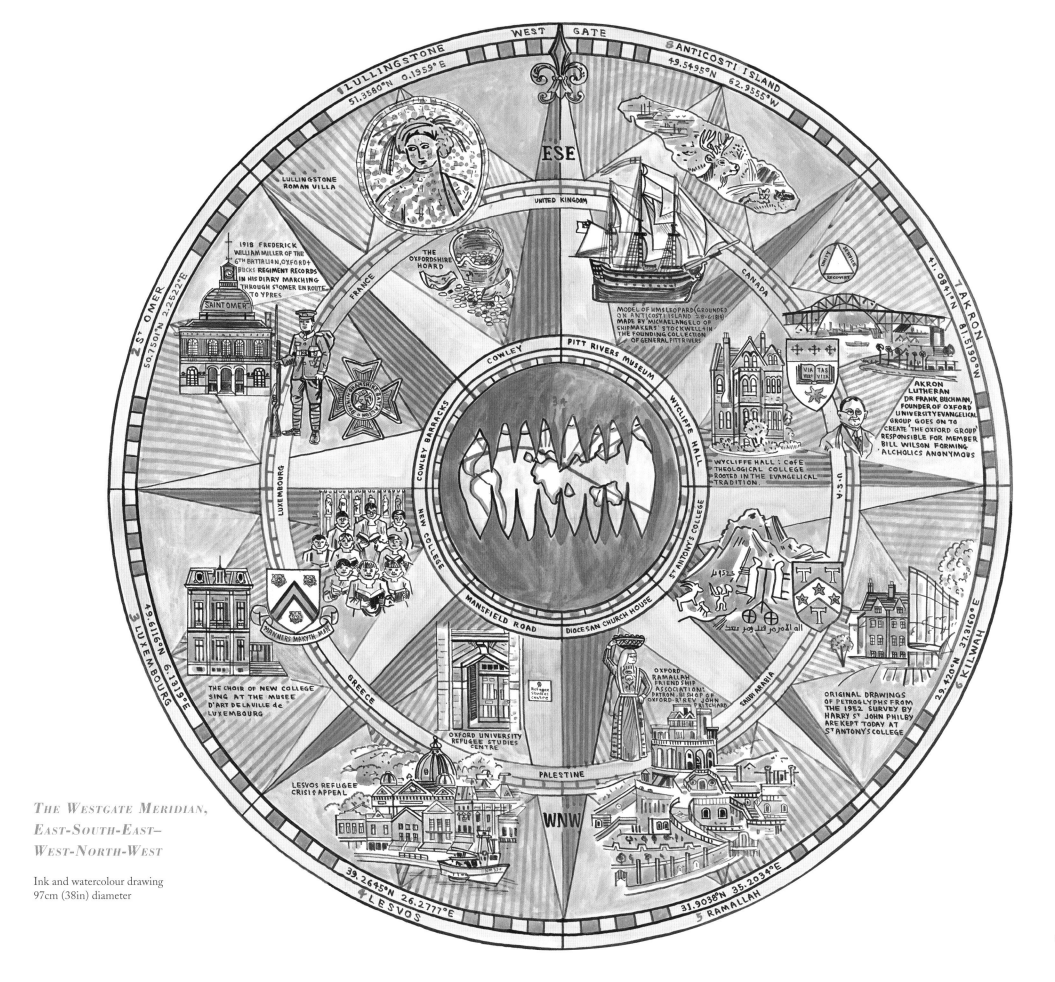

The Westgate Meridian,
East-South-East—
West-North-West

Ink and watercolour drawing
97cm (38in) diameter

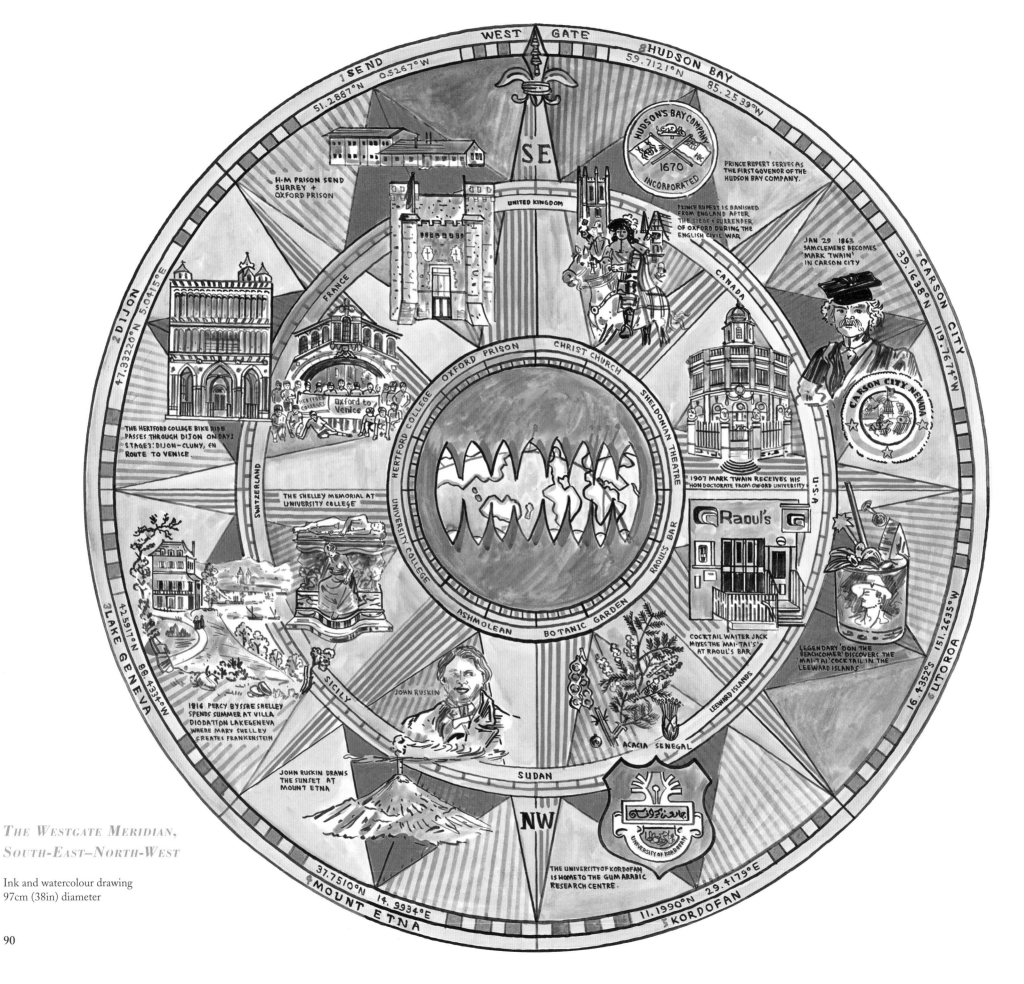

THE WESTGATE MERIDIAN,
SOUTH-EAST–NORTH-WEST

Ink and watercolour drawing
97cm (38in) diameter

90

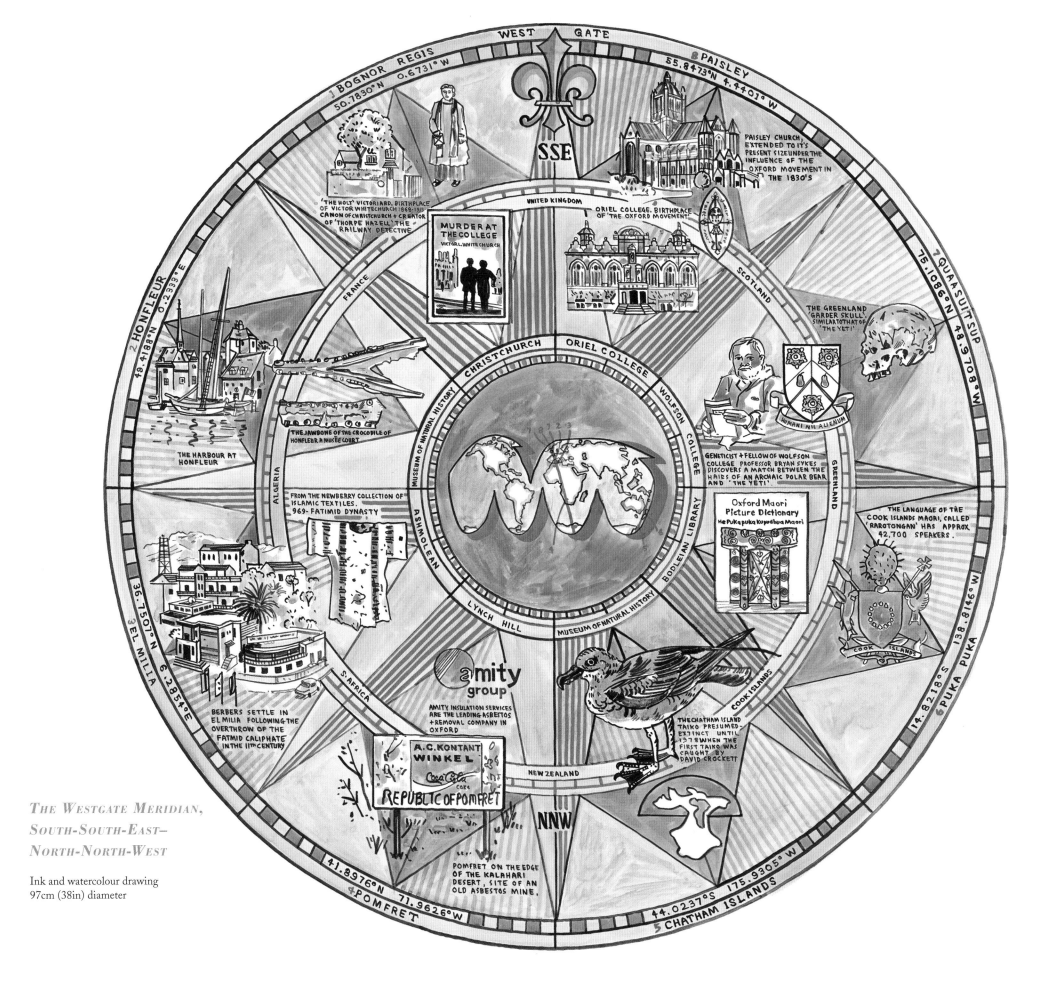

The Westgate Meridian,
South-South-East–
North-North-West

Ink and watercolour drawing
97cm (38in) diameter

THE WESTGATE MERIDIAN,
WEST-SOUTH-WEST–
NORTH-EAST-NORTH

Ink and watercolour drawing
97cm (38in) diameter

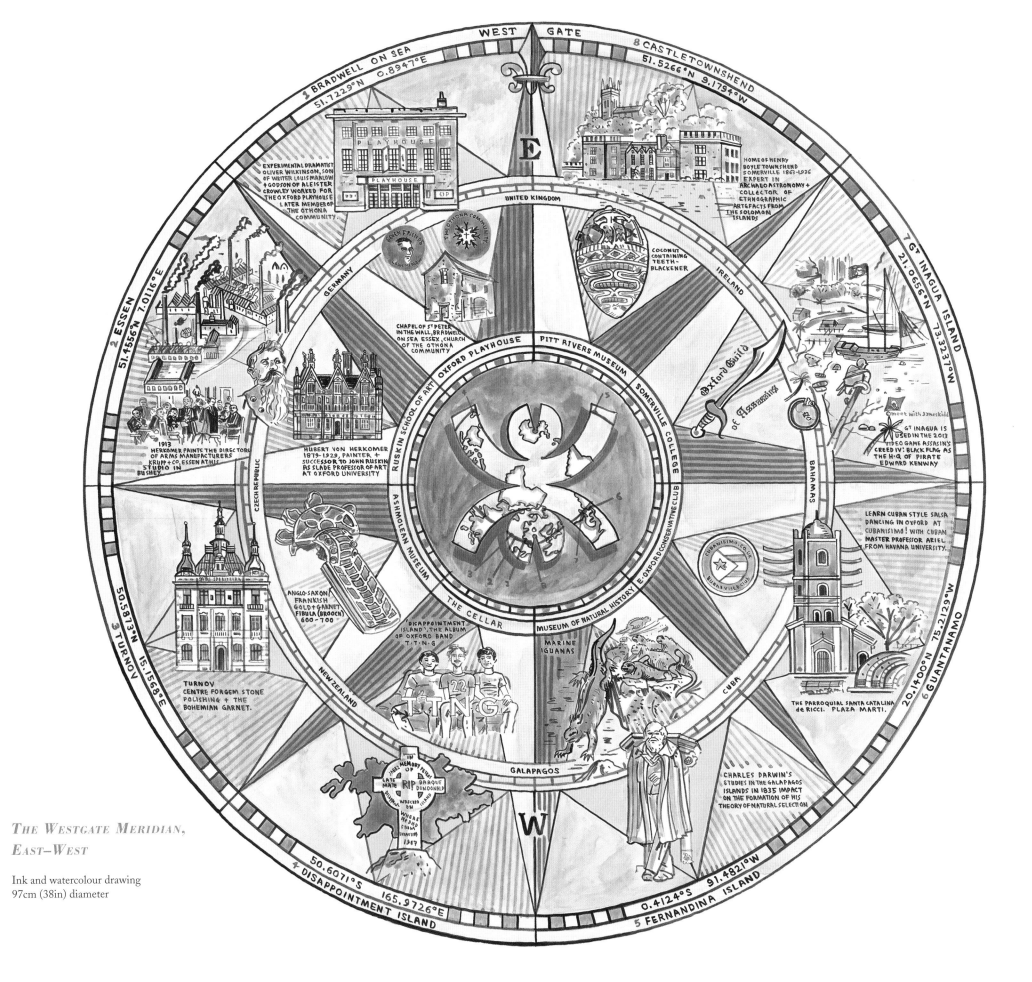

THE WESTGATE MERIDIAN,
EAST–WEST

Ink and watercolour drawing
97cm (38in) diameter

METAPHYSICAL CAMBRIDGE

The metaphysical cyclist proposes a tour of Cambridge according to its sculpture, by bicycle. This tour is a metaphysical one, as this particular way of looking at things allows the terrain of Cambridge, the objects encountered en route and the means of carriage, the bicycle, to exist as a distinct realm. Otherwise it would be just another sculpture tour.

Metaphysics, being a form of study that focuses beyond the physical world, allows the 'sculptures' encountered en route an autonomy as 'things', a 'being' according to how they are encountered, and a 'totality' according to the means by which they are all strung together (as a bicycle tour).

The bicycle has been chosen for this exercise as it represents the consummate metaphysical sculptural object. We know what it is when it is not in use but nonetheless it still looks ridiculous. When we are mounted on it and holding on to the horns of these two-dimensional metal and rubber 'bulls' we become part of the life that we physically and notionally give to objects in their animation.

When we ask if the bicycle is the same thing that it was before we were 'on it', is this any different from asking if a sculpture was a different 'thing' before we were looking at it (metaphysically speaking)? When we are on the bicycle and watching the front wheel revolve before us, do we continue to bear in mind the action and presence of the rear wheel?

Despite its being unseen, its function determined by the properties of 'bicycle-ness' might provide the metaphysical cyclist with an appropriate 'frame' within which to examine sculpture as part of the 'metaphysical' terrain of Cambridge.

In understanding such an exercise in such a fashion we might find it useful to ask the following questions:

1. What is the function, purpose and being of a sculpture when it is not being looked at? Is it the same as that of a bicycle which is not being ridden?

2. Do sculptures that look like us (figurative) possess a different kind of 'thing-ness' than those that don't?

3. Are some sculptures merely fancy, inert vessels for one 'smart-arse' idea or another?

4. What good can come from viewing a sculpture that was created as a memorial within any other context than that for which it was created?

5. What is the point of ascribing 'sculptural values' to things that have not been created as sculptures but which, other than being seen as 'ready-mades', conform to such tenets?

THE METAPHYSICAL CYCLIST: A TOUR OF CAMBRIDGE

Printed pamphlet and hand-tinted print
56 × 76cm (22 × 30in)

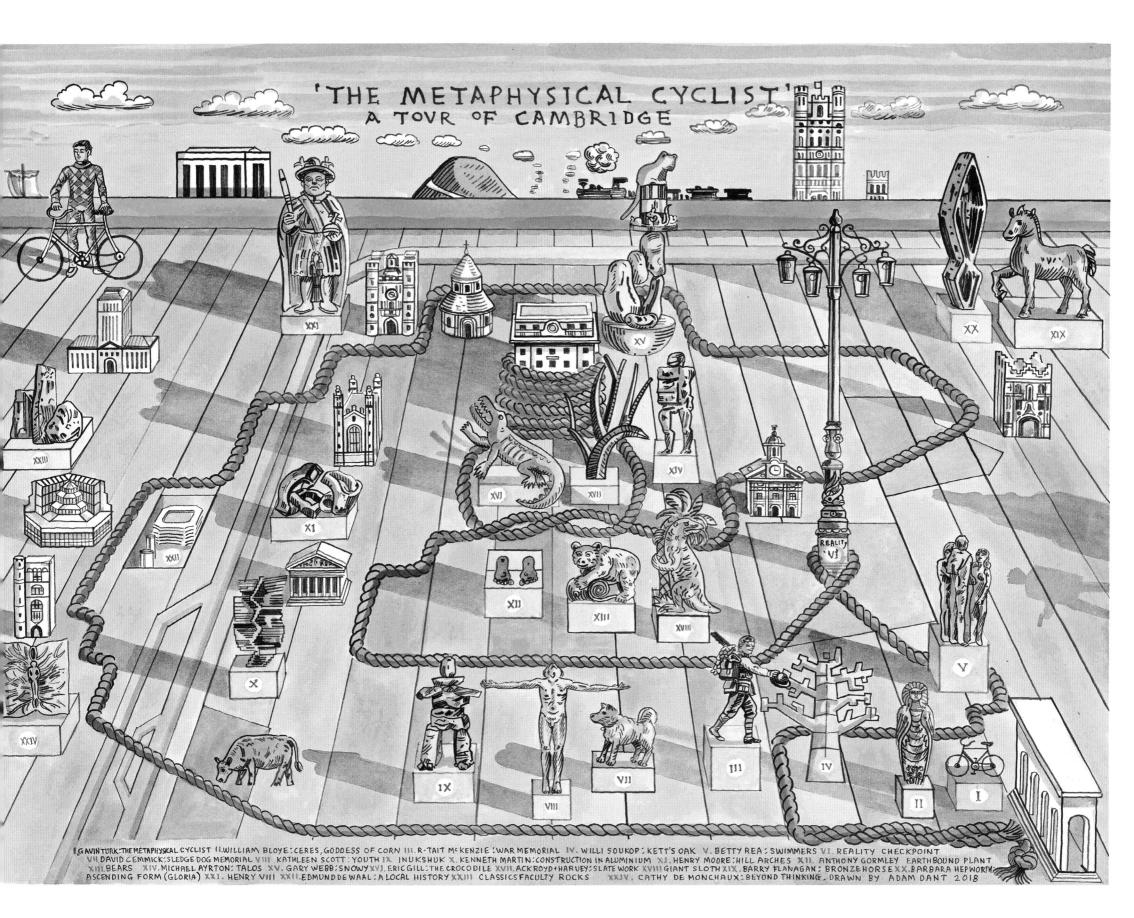

'THE METAPHYSICAL CYCLIST'
A TOUR OF CAMBRIDGE

I. GAVIN TURK: THE METAPHYSICAL CYCLIST II. WILLIAM BLOYE: CERES, GODDESS OF CORN III. R. TAIT McKENZIE: WAR MEMORIAL IV. WILLI SOUKOP: KETT'S OAK V. BETTY REA: SWIMMERS VI. REALITY CHECKPOINT
VII. DAVID CEMMICK: SLEDGE DOG MEMORIAL VIII. KATHLEEN SCOTT: YOUTH IX. INUKSHUK X. KENNETH MARTIN: CONSTRUCTION IN ALUMINIUM XI. HENRY MOORE: HILL ARCHES XII. ANTHONY GORMLEY: EARTHBOUND PLANT
XIII. BEARS XIV. MICHAEL AYRTON: TALOS XV. GARY WEBB: SNOWY XVI. ERIC GILL: THE CROCODILE XVII. ACKROYD + HARVEY: SLATE WORK XVIII. GIANT SLOTH XIX. BARRY FLANAGAN: BRONZE HORSE XX. BARBARA HEPWORTH:
ASCENDING FORM (GLORIA) XXI. HENRY VIII XXII. EDMUND DE WAAL: A LOCAL HISTORY XXIII. CLASSICS FACULTY ROCKS XXIV. CATHY DE MONCHAUX: BEYOND THINKING. DRAWN BY ADAM DANT 2018

THEATRELAND

During the Great Plague of 1665 the first places to close their doors and abandon their performances mid-run were London's playhouses. This infelicitous aspect of the capital's theatrical history was to be repeated in 2020 as a new pandemic for the 21st century forced the powers that be to shut the theatres, the racetracks, the sports stadia and the bingo halls, not just in London but across the country and the globe. Real life quashed the realm of the fantastical once more.

As silences lengthened within these shuttered venues, the sound of laughter, awe and applause diminishing with each passing evening, the stalls unoccupied and the only mousetrap on stage being to catch the new residents, we could imagine the shy spectres of Theatreland's illustrious past being lured back on to ghostly, empty stages to relive past glories in the abandoned auditoria of the Apollo, the Lyric and the Adelphi.

Could that odd mysterious creak from boards untrod for months mark the entrance once more from the wings of Olivier's Richard III? And is what looks like a wisp of smoke rising from the orchestra pit emanating from the embers at the end of a cigarette holder in the elegantly unfurled fingers of Noel Coward? We would like to believe that – like theatre itself never dying – great performances live forever.

A map of London's Theatreland during this unusual health-hiatus, while at the time being of practical use to the student of playhouse facade architecture or the aficionado of the fading, peeling playbill, is really meant as a backdrop to the craft and practice of the many individuals depicted across the West End next to the venues of their songs, dances and declamations. The legendary status and place in the common consciousness of creations such as Mary Poppins, Billy Elliott, and Vladimir and Estragon will be familiar to even the most irregular of theatregoers. Familiar figures from thespian history waft across the cityscape of Theatreland as the likes of Sarah Siddons, Dan Leno, Henry Irving and Richard D'Oyly Carte can all be found waiting for the curtain to rise again, for the audience to settle to a reverential hush and for the spoon to be taken out from under the lid.

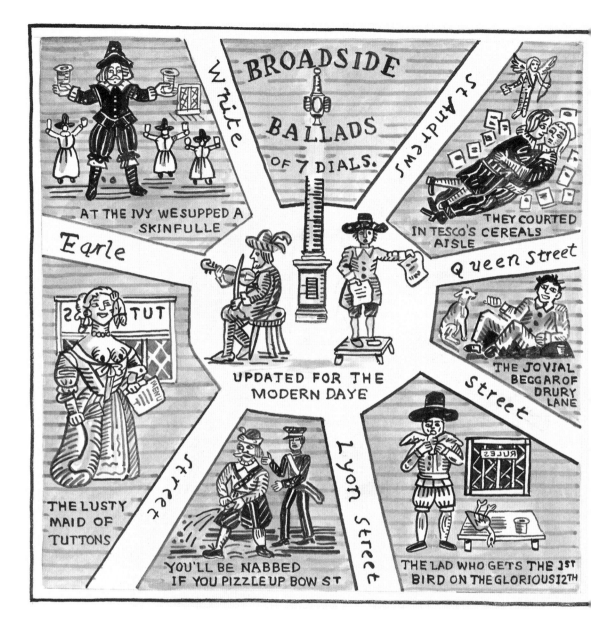

BROADSIDE BALLADS
OF SEVEN DIALS

Ink and watercolour drawing
42 × 42cm (16½ × 16½in)

THEATRELAND

Hand-tinted print
56 × 76cm (22 × 30in)

A Map of the National Gallery: London's Collection of Subliminal Images

Subliminal images, sometimes known more generally as pareidolia, are supposedly the result of the mind's tendency to try to impose a meaningful interpretation onto random visual phenomena.

Finding faces, objects and patterns in the appearance of the natural world has, throughout the ages, moved beyond the realms of being an unwelcome diversion or an activity for the casual amusement of the characters in a *Peanuts* cartoon, to being a bona fide method for the construction of networks that have the capability of encompassing and incorporating via interpretation a deeper understanding of both the natural world and the processes by which it is perceived and categorized philosophically.

The application of the 'pareidolic tendency', though often conveniently sidelined, is hardly insignificant. Perhaps the most significant example is that of the night sky, whereby even at the dawn of human consciousness there is evidence in the form of prehistoric mark-making of this random stellar display being formulated into a vestigial astrological menagerie of ideograms and symbols. The names of the constellations are forever intrinsically linked to the stories of the gods and the ancients, even for astronomers searching for evidence of the birth of the universe.

The serious work undertaken in the late 1990s by the Bureau for the Investigation of the Subliminal Image (the BISI, itself a department of The London Institute of '-Pataphysics) in the realm of the public art gallery sought to reveal a vast trove of imagery that had for too long remained hidden from gallery visitors and curators alike.

Through a carefully structured process and via the application of novel art-historical and psycho-analytical orthodoxies, the BISI's initial catalogue of subliminal images in the collection of the National Gallery, London, provided a fresh insight into the subconscious realm of the world's most important works of art while mapping out a floorplan of the museum's own subconscious presence.

SUBLIMINAL IMAGES IN THE COLLECTION OF THE NATIONAL GALLERY, LONDON

Sepia ink drawing
56 × 112cm (22 × 44in)

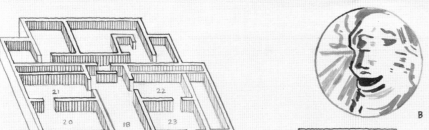

NATIONAL GALLERY LONDON.
BUREAU FOR THE INVESTIGATION OF THE SUBLIMINAL IMAGE

B·I·S·I

A Leonardo da Vinci — The Virgin of the Rocks
B Michaelangelo — The Manchester Madonna
C Raphael — The Alba Madonna
D Titian — Noli Me Tangere
E Giorgione — Il Tramonto
F Botticelli — Adoration of the Kings
G Piero Della Francesca — The Baptism of Christ
H Claude Monet — Gare St Lazare
I Vincent Van Gogh — Chair with Pipe
J Albrecht Durer — St Jerome
K Piero Della Francesca — Nativity
L Mantegna — Agony in the Garden
M Jan van Eyck — The Arnolfini portrait
N Giovanni Bellini — Agony in the Garden

99

About the B.I.S.I

The Bureau for the Investigation of the subliminal image through questioning suggested forms as they appear regularly in the museum, gallery, library, press, decorative *Environment* and advertising continues to establish a validity for overlooked means of visual communication that exist beyond both the subjective subconcious and the historically prescribed.

©B·I·S·I 2001

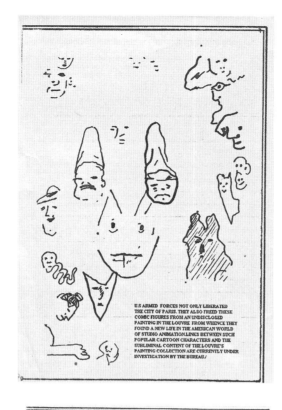

U.S ARMED FORCES NOT ONLY LIBERATED THE CITY OF PARIS. THEY ALSO FREED THESE COMIC FIGURES FROM AN UNDISCLOSED PAINTING IN THE LOUVRE FROM WHENCE THEY FOUND A NEW LIFE IN THE AMERICAN WORLD OF STUDIO ANIMATION.LINKS BETWEEN SUCH POPULAR CARTOON CHARACTERS AND THE SUBLIMINAL CONTENT OF THE LOUVRE'S PAINTING COLLECTION ARE CURRENTLY UNDER INVESTIGATION BY THE BUREAU/

LOOKING AT PICTURES
IN A MANNER PRESCRIBED BY MUSEUMS

THE APPLICATION OF THE SELF HELP GUIDE TO THE COLLECTION BY CURATORS FOR THE PUBLIC GOOD.

THE VIEWING AND APPRECIATION OF PAINTINGS, REGARDLESS OF STYLE, ERA, AUTHOR ETC ETC TAKES THREE PRIMARY OPERATIONAL FORMS. THESE ARE........

...SCANNING AND GLANCING
...SELECTIVE APPRECIATION
...PARALLEL VIEWING

THE BUREAU FOR THE INVESTIGATION OF THE SUBLIMINAL IMAGE THROUGH THE PUBLICATION OF A CATALOGUE TO THE SUBLIMINAL IMAGES IN THE POSSESSION OF THE MUSEE DU LOUVRE SEEKS TO PROVIDE VISITORS WITH THE MEANS AND METHOD TO ALLOW FOR A SUBLIMINAL VIEWING AND APPRECIATION OF THE COLLECTION CAPABLE OF AUGMENTING EACH OF THE OPERATIONAL FORMS LISTED ABOVE.

BUREAU FOR THE INVESTIGATION OF THE SUBLIMINAL IMAGE

A BRIEF INTRODUCTION TO THE CATALOGUING OF SUBLIMINAL IMAGES IN THE COLLECTION OF THE MUSEE DU LOUVRE, PARIS
B.I.S.I
2001

"carrying the torch for a tradition of locating things in the shadows cast by the flickering of said."

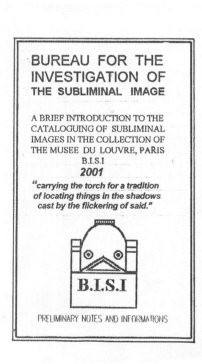

B.I.S.I

PRELIMINARY NOTES AND INFORMATIONS

SCANNING AND GLANCING
THE EYE OF THE VIEWER OPERATES IN AN INDISCRIMINATE OR SEEMINGLY RANDOM MANNER, EXPERIENCING THE MUSEUMS COLLECTION THROUGH MANY BRIEF AND RANDOM ENCOUNTERS.

SELECTIVE APPRECIATION
additional knowledge such as author/subject/era/geography etc etc is applied to and augments the viewing process (e.g reading the label, comparison with other works).

PARALLEL VIEWING
RECOURSE TO INFORMATION BOTH SPECIFIC AND GENERAL ABOUT THE WORK IN QUESTION CONSTITUTES THE MAIN FACTOR IN THIS OPERATIONAL FORM OF VIEWING. PARALLEL VIEWING IMPLYS THAT DIRECT ENCOUNTER WITH THE WORK IS PAIRED WITH IF NOT DOMINATED BY SOME KIND OF PARALLEL RELATIONSHIP WITH THE WORK THAT TAKES PLACE IN A SPACE APART FROM THE WORK AND QUITE OFTEN PERTAINS TO MATTERS OF INDIRECT RELEVANCE. (E.G PERSONAL MATTERS SUCH AS THE COMPLETION OF STUDY AND AQUIRAL OF QUALIFICATIONS , VARIOUS SORTS OF INDIVIDUAL ' ARTISTIC PROJECTS' ET AL).

HELPFUL advice

before undertaking any first-hand encounter with any of the pictures in the museums' collection, THE VISITOR will find it useful to first of all position his or herself comfortably in the centre of a busy gallery in the museum and from this vantage point, to observe in the behaviour of other visitors an active demonstration of the three forms of encounter described above.

i) NOSE   ii) GARDENER   iv) MOLE   v) CHICKEN.

BUREAU FOR THE INVESTIGATION OF THE SUBLIMINAL IMAGE

Recent investigations by the bureau have focused on French Paintings in the collection of the musee du Louvre .These have been conducted with the aim of producing a comprehensive guide to the Subliminal images currently in the possession of the French State.

A validity for the paintings as a collection based on their subliminal visual content has taken many forms over the last few months, including 1. The recurrance of The Frog / Toad motif. 2. A comparison of the subliminal Skull in French Painting with that manifested in Italian religeous painting. 3.Familiar comic characters in the draperies of classically themed works. 4.The anticipation of technological discoveries as witnessed in the skies of Poussin et al.

The Bureau in seeking such validity is thus willing to investigate sightings of subliminal images ,in this instance within French Paintings, made by visitors to the museum, and to draw up reports of such.

Specially commissioned equiptment designed to locate subliminal imagery is proving effective in the assesment of such public submissions.

i) SKULL   ii) VULTURE   iii) TYPIST   iv) PROFILE

WITH COMPLIMENTS

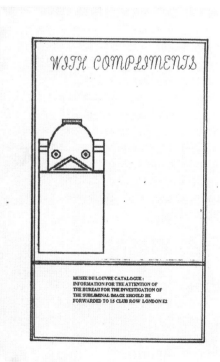

MUSEE DU LOUVRE CATALOGUE : INFORMATION FOR THE ATTENTION OF THE BUREAU FOR THE INVESTIGATION OF THE SUBLIMINAL IMAGE SHOULD BE FORWARDED TO 15 CLUB ROW LONDON E2

Pamphlet explaining the methodology of the B.I.S.I.

13. Graphic processing of 'subliminal images'.

14. The Notion of Subliminal Research.

1. What is meant by 'Subliminal Images'?

2. From Sub-Conscious to Icon.

15. Calculated documentation of the subliminal.

16. 'A guide to the museum's collection of 'subliminal images'.

3. Creative Perception.

4. An Imitation of Life.

THE BUREAU FOR THE INVESTIGATION OF THE SUBLIMINAL IMAGE

A CATALOGUE OF SUBLIMINAL IMAGES IN THE COLLECTION OF THE MUSEUM

Within the fabric of the Museum's collection of paintings exist a vast number of well known as well as undiscovered 'hidden' or 'subliminal' images.

A famous case in point was discussed by Sigmund Freud. Referring to Leonardo da Vinci's 'Madonna and St Anne' and the painting's concealment of a vulture, Freud sought to speculate on the sub-conscious mind of the artist. An analysis of the sub-conscious or subjective viewing of such paintings often reveals that the appearance of such 'sub-visual information' (subliminal images) in a similar spirit can assist in the interpretation of particular artists works as well as providing curious and previously overlooked routes through the collection as a whole.

The rag-bag collection of a single days visitors to the Museum and the reviewing of the paintings collectively has parallels to the randomness of the paintings in the museum itself when structured as a collection.

'The Bureau for the Investigation of the Subliminal Image' highlights the presence of the museum as a sub-conscious historical entity. It's presence as such is identified through the often unpremeditated and subjective viewing processes of it's visitors.

This pamphlet provides background information and a description of the working methods used, to enable the production of a comprehensive guide to the museum's collection of subliminal images.

5. The Museum at night.

6. The Museum as a sub-conscious realm.

9. Museum Traffic.

10. Introducing the 'Subliminal Viewing'.

7. How we look at paintings.

8. The cult of the visit.

11. Unmediated Sightings.

12. Charting Instances of the Subliminal.

Press release for the launch of
*The B.I.S.I.'s Catalogue of
Subliminal Images in the Collection
of the National Gallery, London.*

TULIP FEVER

The black and 'delft blue' print 'Tulip Fever' crams the history of the world's first supposed financial bubble into a traditional Dutch interior in the style of a dense contemporary 'cabinet of curiosities', the curiosities within being: chronological tales of events from the birth, passage and culmination of 'Tulip Fever'; depictions of key individuals from 'Tulip History'; and objects selected from the venues where the original picture was put on display, the Museum Van Loon in Amsterdam and The Bowes Museum in Barnard Castle, County Durham.

A decorative border around the print describes the various commodities traded en route by the globe-trotting vessels of the Dutch East India Company, the new wealth of whose directors led to the prizing of tulip bulbs as status symbols in the first place.

The very particular stories of the participants in the Tulip Mania of the Dutch Golden Age in the early 17th century are told in rhyming form, illustrated as a set of playing cards that slot together to create a pictorial view of a typical Netherlandish knot garden in the 'Game of Flora'.

Just as the original obsessive and 'completist' mania of paying 'top guilder' for tulip bulbs echoes the current phenomenon of speculative investment 'bubbles', Flora is a form of the familiar Happy Families card game, though given the fate of the average tulip speculator at the time possibly better compared to the game of 'Donkey'.

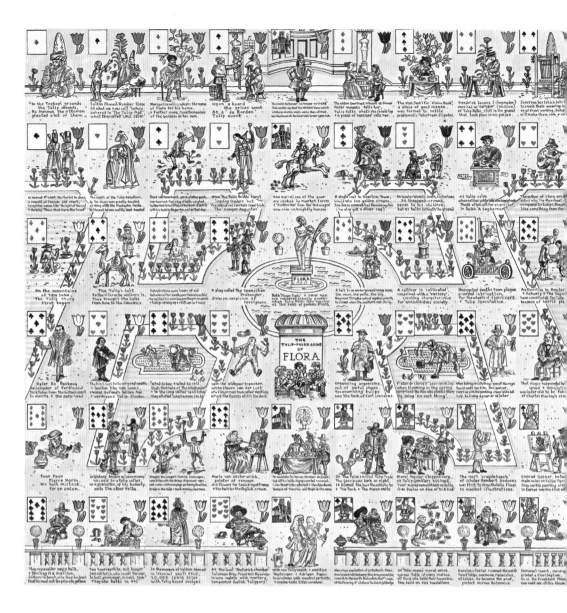

THE GAME OF FLORA

Printed board game
and playing cards
102 × 102cm (40 × 40in)

TULIP FEVER

Two-colour lithograph
58 × 84cm (23 × 33in)

TULIP FEVER

Coco-opolis

At one time hapless or desperate painters who had succumbed to the lure of popular appeal and embarked on a course of churning out inoffensive depictions of pretty sailboats, lemons in chipped rustic bowls and, at the bottom of this particular aesthetic barrel, saucer-eyed fluffy kittens, would have been branded 'chocolate-box artists'.

The opportunity came for me to join the ranks of those who practise what is known in France as 'L'art Pompier' (though that particular ladder awaits) when I was commissioned to create a 'masterpiece' for a famous Birmingham chocolatier.

'Coco-opolis' was the resulting work of art and this huge 'composite' sepia ink map/panorama, as well as adorning the lobby of the chocolate factory, was also broken into segments, much in the fashion of a fruit and nut bar, each of which adorned a different box of fondant delights. When placed together these six chocolate-box paintings formed a complete coco-landscape in whatever order they were combined to tell the story of the history of chocolate.

From the botanical significance of *Theobroma cacao*, its use in the rituals and the economies of Mesoamerican culture, through its discovery and 'creolization' via the Spanish Conquistadors, its arrival at the courts of Europe, adoption by the Georgian chattering classes, refinement in the Alps and industrialization for mass consumption by the Victorians, chocolate has a lot to offer as a toothsome core that might be enrobed with the story of man.

The best book on the subject remains Sophie and Michael Coe's *The True History of Chocolate* which, were the viewer in need of a salient and scholarly tour guide to 'Coco-opolis', handily underpins the visual narrative experience with a sound linear chronology.

Coco-opolis

Sepia ink drawing
2.13 × 2.74m (7 × 9ft)

1. The Food of the Gods

The binomial *Theobroma cacao* or 'food of the gods' was given to the cacao tree by Carl Linnaeus in 1753. Difficult to grow outside of areas 20 degrees north and south of the equator, the tree demands year-round moisture and is subject to a multitude of diseases. The seeds or cacao 'nibs' are contained in pods surrounded by a sweet pulp. They are fermented, dried, toasted and winnowed to make cocoa powder.

In 1900, Dr Henri Roberts proposed that the caffeine, theobromine, serotonin and phenylethylamine found in chocolate make it a perfect tonic to be prescribed as an antidepressant and an anti-stress agent. There may also be proof to the doctor's claim that chocolate has an aphrodisiac effect. The Marquis de Sade had a true passion for chocolate, frequently requesting deliveries to the fortress where he was locked up of 'cake black inside from chocolate as the devil's ass is black from smoke'.

2. Mayan and Aztec 'Happie-Money'

Chocolate as it was enjoyed by the ancient Mesoamerican societies was first known as *chcahuatl* by the Mayan people, who believed it to be a gift from the gods. They ground cocoa beans and mixed the powder with water to create a bitter, frothing drink that was often spiced with chilli. The Aztecs, who called it *xocolatl*, revered and valued cocoa as a sacred beverage to be given to victorious warriors after battle and as a currency more valuable than gold.

3. Conquistadors and the 'Creolization' of Chocolate

In the 16th century Hernán Cortés, having established colonies in the New World, returned to Spain with the first cocoa beans to arrive in Europe. Initially used as a medicine, its bitter taste was soon masked with the addition of sugar, honey or vanilla, which saw it become a favourite of the Spanish court.

4. The European Chocolate Craze

The taste for chocolate soon spread throughout the aristocratic families of 18th-century 'Age of Enlightenment' Europe, but despite the sensational status of the novel beverage, the process for the preparation of cacao as described in Diderot's encyclopaedia is hardly changed from the time of the Olmecs.

In Georgian England political and cultural life found a lively and convivial home in the 'chocolate houses'. One of the oldest, The Cocoa Tree in St James's, became a well-known meeting place for Jacobites, and White's Club, which still has its home on St James's Street, had its origins as a chocolate shop founded in 1693 by Francesco Bianco.

5. Alpine Chocolate Ingenuity

Until 1828 chocolate had only ever been consumed as a drink made with water. The cocoa press put into service by the Dutch inventor Coenraad van Houten allowed for the separation of the fat from the cacao bean and for the mass production of a more palatable powder to which milk could be added.

The revolution that still sees Switzerland at the top of the chocolate mountain (Swiss citizens are reported to consume an annual per capita 11lb (5kg) of chocolate) began in 1819 when François-Louis Cailler opened the first Swiss chocolate factory on Lake Geneva to produce chocolate using his own inventions.

The first Swiss chocolate bars were created by Henri Nestlé, who had formulated powdered milk, and Daniel Peter, who came up with the ingenious idea of combining this with cacao powder and replacing the moisture with cacao butter so the mix could be poured into a mould.

Further innovations such as 'conching' in 1879 by Rudolphe Lindt in his production of a smoother 'fondant' chocolate, and the process of 'tempering' to eliminate granular crystals pioneered by Jean Tobler and Philippe Suchard, saw the Swiss gain the reputation and the market they continue to occupy to this day as the world centre of chocolate.

6. The Modern British Chocolate Industry

With the purchase by Joseph Fry in 1789 of a Watt steam engine to grind cacao beans the industrial age shifted chocolate production into a higher gear. Mechanization and a contract to supply cocoa to the Royal Navy soon made Fry's the largest chocolate manufacturer in the world, eventually seeing the company invent the Easter egg in 1873.

Fellow Quaker John Cadbury proved a great chocolate rival to Fry by securing the patronage of Queen Victoria and by importing a model of Van Houten's pioneering *melangeur* mixing machine. As residents of his famous 'chocolate utopia' Bournville village, Cadbury's employees also benefitted from all manner of sporting, cultural and leisure activities, setting a standard that was to be followed by the town of Hershey in the USA. Cadbury is also responsible for introducing the first 'chocolate box'.

SCOTTISH LONDON

'Scottish London' salutes Caledonian connections with the capital, historical and contemporary, depicting a venerable host of 'Cockney Jocks' (as well the odd 'Jockney Cock').

Reluctant to acknowledge the legitimacy of English supremacy as long ago as the 12th century, legend has it that Scottish nobility trucked cartloads of peaty soil down from the Highlands to their diplomatic base at 'Great Scotland Yard' so that their palace wouldn't have to sit on English ground.

The distinctiveness of London's Scottish presence persists on this cartographic compilation. Playgrounds and palaces where a Scot in London can enjoy creating, carousing, politicking and training how to fight people are all pointed out alongside depictions of famous immigrant Scots saluting a convivial yet uneasy union through the ages.

As the casual tourists to London unfortunate enough to find themselves in the West End whenever Wembley Stadium hosts an England–Scotland football match might attest, fears of a Jacobite invasion, such as those depicted in Hogarth's painting *The March of the Guards to Finchley*, are well founded. To witness the pubs and deep-fried pizza joints of Leicester Square as well as all the back alleys and vennels of Covent Garden and Soho besieged by saltire-draped laddies on the lager is a sight to behold (from a safe distance).

Aside from such metropolitan gatherings of the clans, the Scottish contribution to London via the realms of the sciences, arts and engineering, as well as politically and economically, is significant. Stand in awe of the work of Robert Adam, Scottish neoclassical architect in Portman Square, or appreciate the wit of numerous public artworks by Bruce McLean, and you'll soon realize how much the capital owes to the Caledonian spirit. The name of William Arrol, responsible for the construction of Tower Bridge (in Glasgow, before shipping it down to the Thames), and those of his Scottish crane drivers and rivetters can be found studded in brass into its walkway.

The map of 'Scottish London' also features the furiously pacing figure of Malcolm Tucker, the fictitious Westminster monster and 'master of spin' from the political drama *The Thick of It*, whose famously crude and acerbic ejaculations are apparently not so far from the truth in the real Westminster village. I hope that, like his forebears, the soil on which he paces is Scottish soil.

Hand-tinted print
56 × 76cm (22 × 30in)

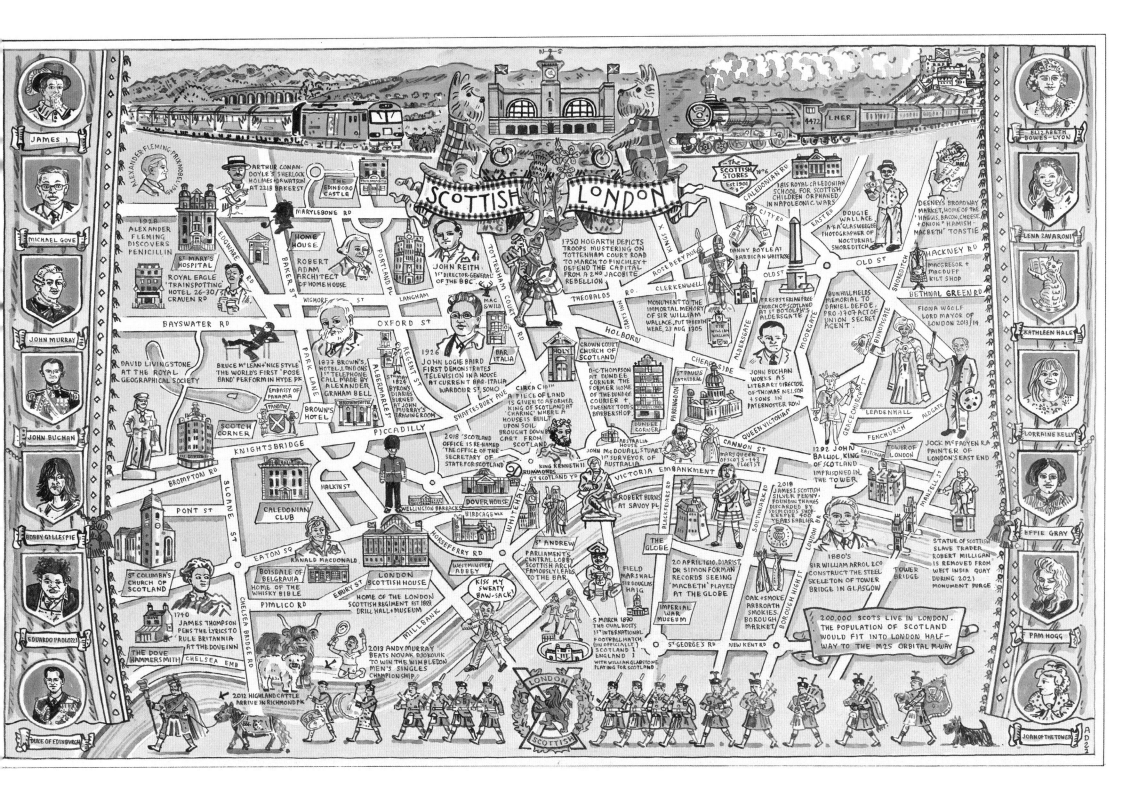

A Walk Round the End of the World

A Description by the Creator of This Epic Walk, Tom King

'The unfashionable end of the Thames.'

'A dirty old man.'

'Britain's most massive backwater.'

These were some of the more respectful phrases used to describe the Thames Estuary back in 2001, the year the first edition of *Thames Estuary Trail* saw the light of day. The introductory chapter described the riches to rags progress of a journey down the Thames, from idyllic Cotswold stream, to royal and capital river, to Skid Row.

Twenty years on, there has been, in the most literal sense, a sea change. The core theme of the first edition – neglect – is now insupportable. Everyone now loves the Estuary. Nobody any longer ignores it.

The resurgence has taken many forms, including, now, this second, 20th-anniversary edition of *Thames Estuary Trail*. It has been commissioned by the Southend-based arts organization Metal, under the auspices of the Thames Estuary Growth Board.

The arts generally have proved a key driver in boosting the Estuary's fortunes. The basis all along has been that culture and regeneration are near synonyms. The launch of this new edition takes place as part of the 2021 Thames Estuary Festival, celebrating and demonstrating the way in which the Estuary has become a hub of creativity.

That process can be viewed in microcosm at High House, in Purfleet. As its name suggests, this old farmhouse occupies a peach of a spot, on high ground, overlooking the Estuary. Yet its 17th-century architects constructed it so that its back is turned on the Thames. Instead, its picture windows choose to overlook the farm's pig-yards. Now, though, High House is a key base for the Royal Opera House, drawing a new generation of talent from along the shoreline. From pigs to opera – in some ways, that sums up the trajectory of the Thames Estuary.

The Thames Estuary Trail, allowing us all to orbit the lower river close-to, is also part of that transformational process. The 2021 Estuary Festival sees the emergence of a parallel companion to the Thames Estuary, a map for explorers and wanderers. This is something quite new. The only other map dedicated specifically to the Estuary in its own right is for seafarers. Stanford's Nautical Chart of the Thames Estuary is no doubt handy when you are trying to avoid the Shivering Sands, but impractical for walkers and cyclists who don't have a rudder bolted astern. The new map, drawn by the brilliant artist Adam Dant, features the cartographic debut of the Thames Estuary Trail.

The original 83-mile Thames Estuary Trail ended at Warden Point, on Sheppey Island – a truly scary location, where the Estuary is fast eating away at so-called dry land, producing a terrain like the Somme in 1916. At a place like this, you stop thinking in terms of hours, days and years. You convert instead to eras and aeons. At this stage the Thames Estuary appears a truly awesome presence. The book's subtitle, 'A Walk Round the End of the World', met all requirements for accuracy under the Trade Descriptions Act.

That being the case, I also thought that the journey was complete. Where do you go after you have hit the end of the world? But the Estuary keeps you on your toes, and the Trail has now been extended along the north Kent shoreline. In the short winter days of 2020/21 I covered the stretch from Sheppey Island to the North Foreland, the point where the Kent shore dips south and the Thames Estuary turns into the English Channel. The Thames Estuary Trail now weighs in at 107 miles. The book's new chapters describe that extension process.

With Warden Point to the rear, the Estuary began to feel a kinder, rosier place, as well as a much more accessible one. The change over 20 years has been striking. Back in 2000, I had most of the trail to myself. I described how the sheer loneliness of the Estuary meant that even seagulls suffered mental health issues. And that was in midsummer.

Now, despite bitter midwinter conditions, the route heaved with a veritable cavalcade of fellow travellers, of all ages and from a range of different places. Londoners, especially, seem to have rediscovered their old affinity with the Estuary. Huge numbers have made the same discovery that I did 20 years ago. To know the Estuary is to love it. Hopefully the new chapters reflect both the love and the awe.

The other great change over two decades concerns accessibility. The original writing of *Thames Estuary Trail* was galvanized by a simple fact. Nobody ever seemed to have walked around the Estuary, ever. But then, why would anyone choose to walk around what Greenpeace dismissed as a 'turd dump'?

For that matter, how could they? Large sections of the shoreline were blocked, notably at Vange and Mucking, London's waste dumps since time immemorial. Elsewhere, footpaths were overgrown and unused. Brambles had to be cut back. I took to packing a pair of toggle-loppers. Still, it had been a lifelong ambition to blaze a trail, and now the opportunity had arrived.

In 2021, that route has become a doddle. One thing you will not encounter if you set out to tackle the Trail is frustration (apart from the little matter of the absence of bridges over the Medway and Swale, dealt with in chapters 12 and 13). Europe's largest landfill reclamation project has transformed the old metropolitan tips into wildlife parks. New, well-signposted bridleways have been laid down. Along quite a few stretches, the seawall now doubles as a hard-surface, bramble-free path.

One thing has not changed. The Estuary itself remains a mesmerizing presence along every yard of the Trail. The first chapter describes how something dismissively referred to as a 'body of dirty water' crept up unawares, and dispatched me on a journey. It has gone on doing so ever since. Its waves have lapped away at me, but not in a destructive, Warden Point way. 'Sooner or later, the Estuary will wash us all away,' was the conclusion of the final chapter. For now, it is having a different effect. Its grandeur, and the enchantment of its marshes and hidden creeks, soak away all negative thoughts. It brightens lives, literally, with that famous light beloved of J.M.W. Turner, drawing us onward in an extended walk, a continuing journey.

THE THAMES ESTUARY TRAIL: A WALK ROUND THE END OF THE WORLD

Hand-tinted print
56 × 112cm (22 × 44in)

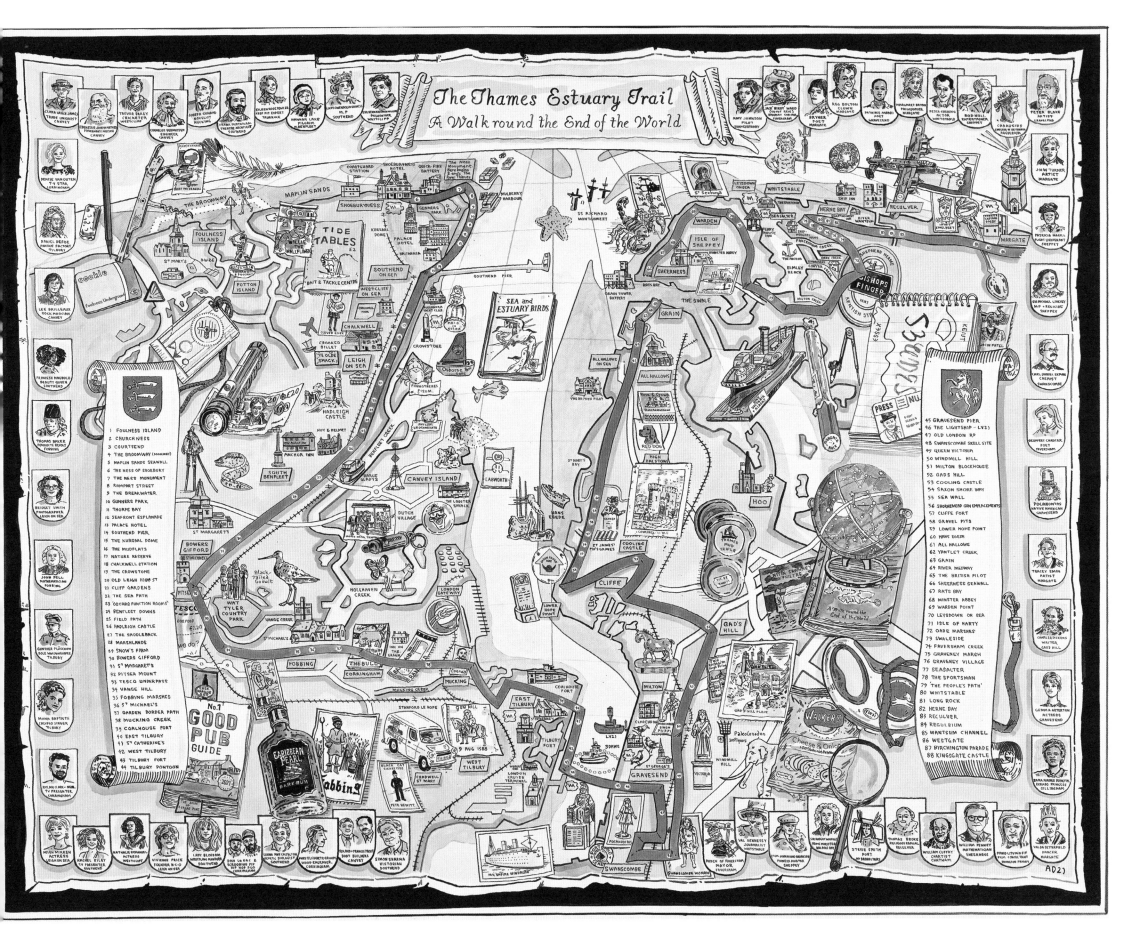

'LINDON': A MAP OF LONDON AS IT MIGHT HAVE BEEN

How many times have we all conjured up the presence of that alternative version of the capital known as 'Lindon'?

A tiny, inadvertent slip on the keyboard from 'o' to 'i' can take us to a realm that's nearly the city of daily trials, tribulations and toils for millions of people among its towers, transport and terrain, but not quite. Familiar landmarks are suddenly replaced by bizarre quasi-counterfeits whereby not just the structures of our own age but edifices from centuries past become cuckoos in the Londoner's nest. It's as if either the conflagrations of the Great Fire and the Blitz had never happened or that a bizarre slippage has occurred to transport us sideways and into a strange counter-factual reality.

All of the architecture and infrastructure of 'Lindon' has emanated from the same drawing boards that fostered the familiar and much-loved civic architecture of the capital. Whether the strange and ersatz backdrop to our imagined existence in this new city, twinned via some form of pataphysical phenomenon, represents a new version of the physical world as an afterthought, of second-best ideas or even of concepts for which the public were not yet ready, the fact is that once all these aborted schemes are depicted across the two sides of the eternal Thames (though the river itself has not been immune to various proposals to drain, straighten and rationalize it) 'Lindon' certainly possesses all of the same dystopian quirks, stupid corners, poorly thought-out street layouts and ugly carbuncles that the GLC and local authority planning officers have foisted on London anyway.

Londoners might laugh at the audacity of the 'Charing Cross Helidrome' or the 'Thames Embankment Sunken Highway', but are such ambitious caprices any worse or short-sighted than building an artificial mountain at Marble Arch, destroying masterpieces of English Victorian classicism such as the Coal Exchange to widen a street that the same corporation only go and narrow again 50 years later, or ripping the insides out of countless characterful buildings to leave the ghosts of the capital's past hovering behind a facile mask that bears the smug grin of a generation who are so much smarter than their predecessors?

Welcome to Lindon.

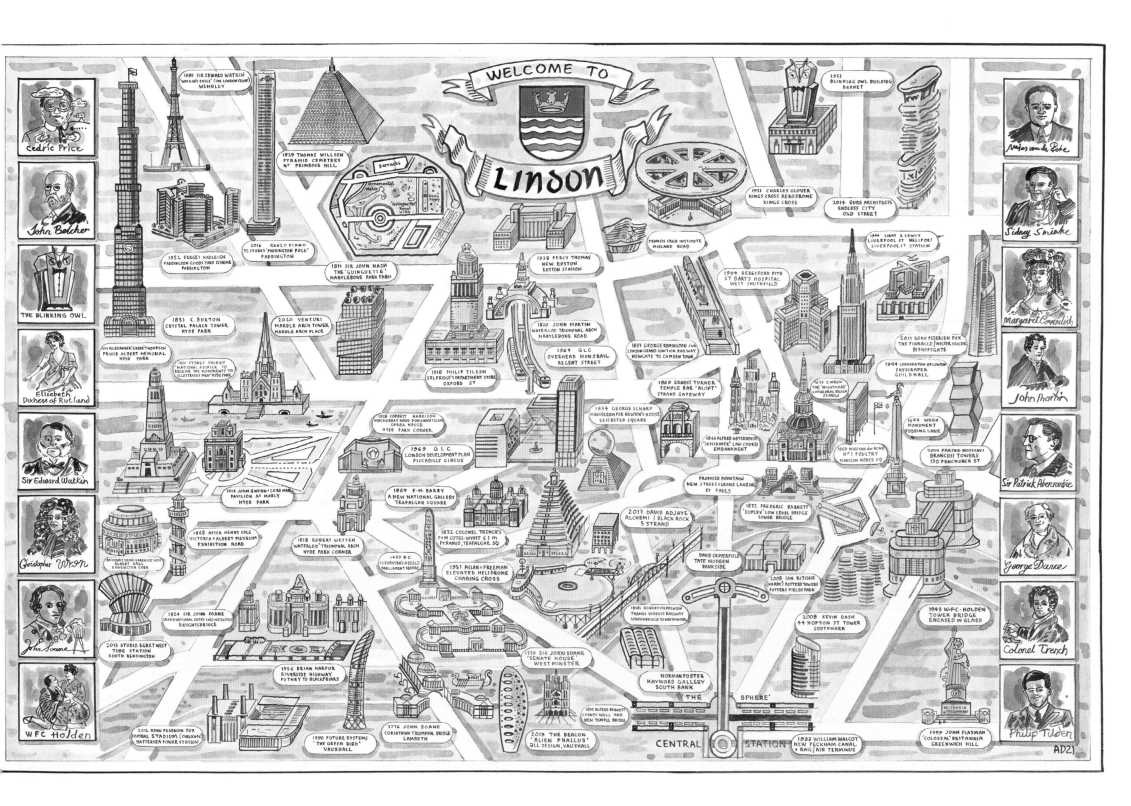

111

THE MIND OF EAST LONDON

London's eastern quarters are often spoken of as offering a new home for successive generations of the inbound. Conversely, the same regions beyond the City of London's walls have been a convenient place to push out the unwanted and undesirable members of a society that can only function according to certain norms.

Even today we can witness what was once known as Bedlam, London's asylum for the mentally ill through the ages, within which the predicament of its inmates was slowly better understood, pushed to the very edge of the present capital where it exists in its current incarnation.

The original hospital of 'St Mary of Bethlehem' was more of a religious refuge than a place for medical treatment. It stood where Liverpool Street station empties out its bustling hordes of City workers during the working week and at the weekend cocktail-clutching seekers of abandon. The world of the troubled mind – for the time being – far from the minds of both.

The kind of trouble caused by raucous actors in the Elizabethan era saw members of the profession forced to worship at their own church outside the city walls too. The tombs of many actors, jesters and dancers of jigs can be found at St Leonard's, Shoreditch. Jacobean playwright Thomas Dekker visited Bedlam regularly, not to gawk at the poor inmates as other visitors had paid to do, but to better draw the scenes in his plays set in the institution. William Shakespeare was assisted by Timothie Bright, the author of *A Treatise on Melancholie*, in his understanding of madness and mania to create characters such as Hamlet and Lear. The 'wandering fool' would have been a common sight to Shakespeare and his ilk.

That the causes of madness in the Elizabethan and Jacobean era were ascribed to excessive wine, comets and demonic possession, placed mental illness within a religious context. It was reported that both audience members and actors at a performance of Marlowe's *Dr Faustus* were sent mad by the appearance of 'actual demons' on stage.

In the Georgian era, John Wesley was not considered mad when he experienced his divine revelatory moment on the City Road where his resulting chapel now stands. He was 'inspired' by a vision that would expand to occupy the souls of tens of thousands of followers. Surely not a case of mass hysteria. The parents of writer Mary Lamb might have thought her conversion to 'muhummadanism' at the age of 12 a bit odd, but it was her stabbing to death of her mother that saw her sent to the Hoxton House Asylum.

The East End as an abode of the desperately poor was famously recorded by Jack London in photographs and in his harrowing 1903 account of life in the slums of Whitechapel, *The People of the Abyss*. Concerned at the plight of the underclasses, Jack London was one of many socialists at the time who believed that the physically and mentally enfeebled Untermensch should be subjected to the new science of eugenics as part of a Marxist social unshackling of the working classes.

The same working poor and their perpetual physical and mental woes would at the same time be subjected to more enlightened programmes of 'betterment', such as through the work of the social reformer Henrietta Barnett who, via provision of alternative routes, education, employment etc., ensured that young girls were able to avoid the inevitable slide into drudge work, prostitution and all the consequent ills.

The 'mentally unstable' and 'neuro-divergent' still live with us on the streets of Spitalfields and Aldgate. Though the 'grafting', crack-addicted beggars of the 21st century might not – as so-called 'sturdy beggars' such as Tom o' Bedlam was – be in possession of a licence to beg, they nonetheless have the support of case workers to oversee the continuing experiment of 'care in the community'.

THE MIND OF EAST LONDON

Hand-tinted print
76 × 112cm (30 × 44in)

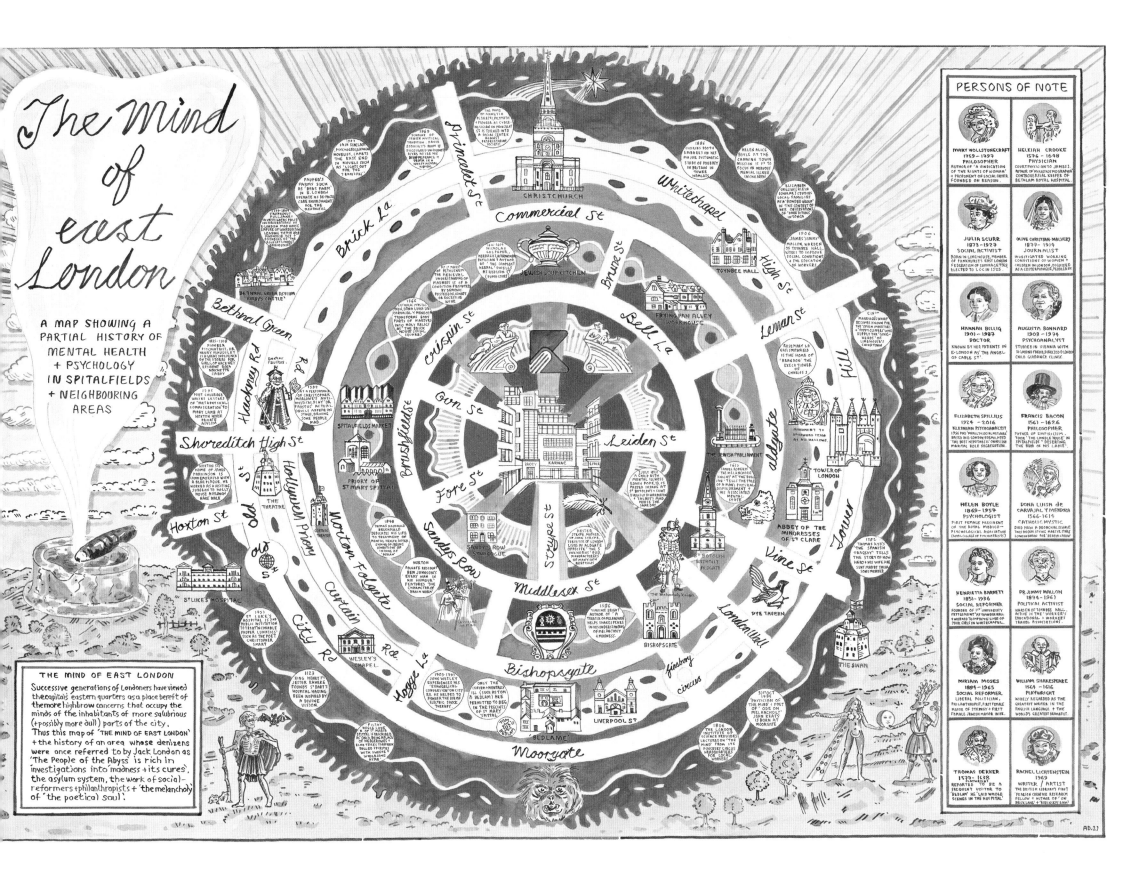

THE GENTLE AUTHOR'S TOUR OF SPITALFIELDS

Celebrated figures from past and present appear across this map of one of London's most historic enclaves, as described in the writings of the pseudonymous Gentle Author of Spitalfields Life. Originally settled by the Romans, the discovery of the tomb of a noble Roman girl buried wearing expensive silk garments coincidentally provides a thread to the Huguenot silk weavers who thrived in the same neighbourhood centuries later. Created to underpin a walking tour of Spitalfields that weaves through the ghost worlds of the Romans, Huguenots, medieval plague victims, Victorian costermongers, members of the Jewish and Bangladeshi diaspora, and the recently dissipated Britart scene, this cartographic encapsulation of a neighbourhood in time updates the previous 'Map of Spitalfields Life' as a living historical terrain.

THE GENTLE AUTHOR'S
TOUR OF SPITALFIELDS

Hand-tinted print
76 × 112cm (30 × 44in)

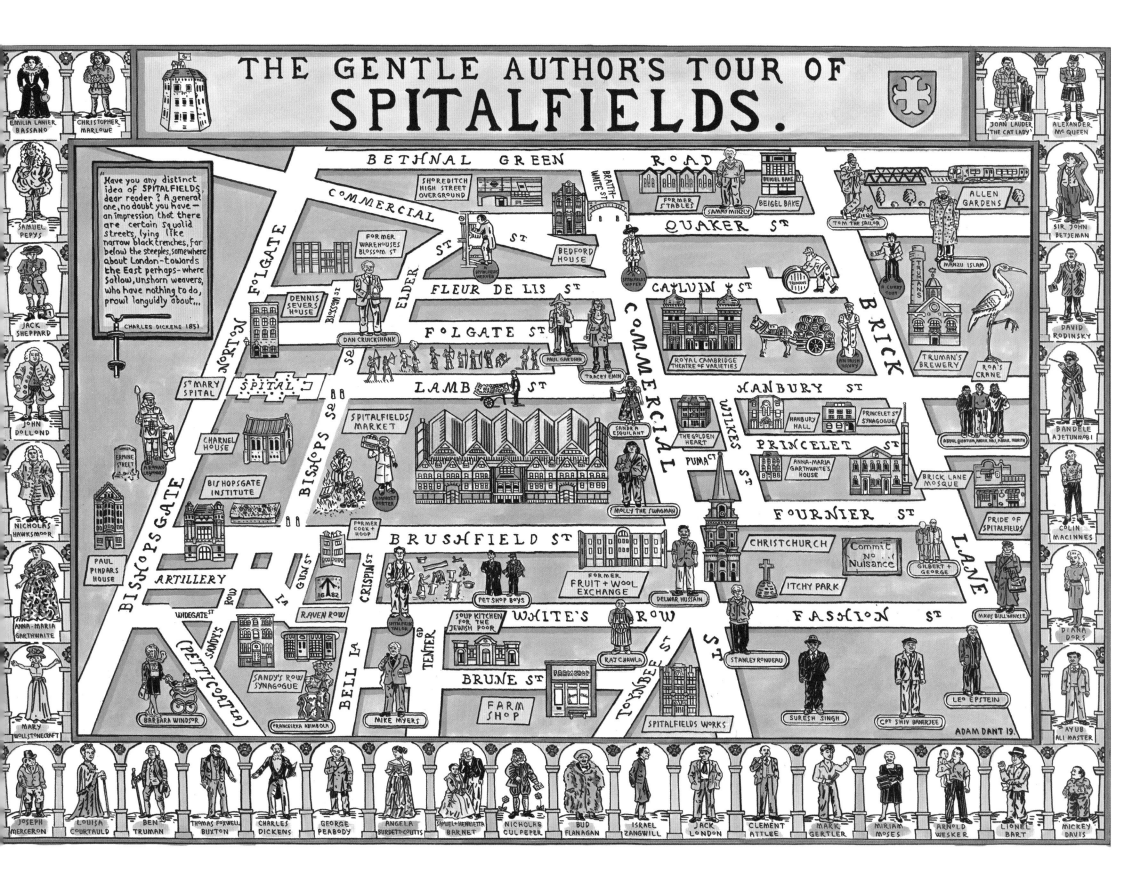

THE GENTLE AUTHOR'S TOUR OF SPITALFIELDS.

"Have you any distinct idea of SPITALFIELDS, dear reader? A general one, no doubt you have — an impression that there are certain squalid streets, lying like narrow black trenches, far below the steeples, somewhere about London — towards the East perhaps — where sallow, unshorn weavers, who have nothing to do, prowl languidly about..."

CHARLES DICKENS 1851

ADAM DANT 19.

One of the earliest uses of the word 'sleazy' appeared in 1644 when the courtier and natural philosopher Sir Kenelm Digby, in his treatise 'Of the Nature of Man's Soule', deployed the term to describe the human brain as being:

'... a substance composed of watry parts mingled with earthy ones: which kind of substance we see are usually full of strings, and so in strong hard beere, and in vinegar, and in other liquors of the like nature, we see (if they be exposed to the sunne) little long flakes, which make an appearance of wormes or maggots floating about. The reason whereof is, that some dry parts of such liquors, are of themselves as it were hairy or sleasy, that is, have little downy part, such as on the legs of flies, or upon caterpillars, or in little locks of wool ...'

Guilty politicians might very well deploy Digby's description of the 'fuzzy' nature of the brain when attempting to explain away just how they found themselves using taxpayers' cash to furnish their garden pond with a luxury duck house, or why they informed the whole nation that their breakfasts were mostly poisoned, or how they ended up *in flagrante* in the bushes with a guardsman after deciding to take a late-night stroll home through the park, or how they came to flee a Department of Trade and Industry investigation into their 'creative accounting' and allegations of spying by flying to Miami, faking their own death, assuming the identity of the deceased husband of a constituent, pitching up in Australia with their mistress/secretary, spending six months trying to get asylum in Mauritius and Sweden on arrest, spending 64 days in court defending themselves and then refusing to quit as an MP while on remand in Brixton Prison.

Not all webs of political deceit are as tawdry as that of John Stonehouse, the subject of the last list of transgressions, though many are just as tangled. Though, for the most part, sex and money appear to be at the heart of the sticky moments depicted across this map of the capital as 'The Paradise of Sleaze', some politicians have found themselves targeted over the most stupid petty affairs by a press hungry for a scoop or keen to embarrass a powerful figure for partisan point-scoring.

One case in point was 'Plebgate', whereby not only did serving police officers confect a spurious allegation in response to the grouchy demeanour of MP Andrew Mitchell, accusing him of using the derogatory term 'pleb' against them, but in backing up their claim they invented a crowd of 'visibly shocked' onlookers as part of their narrative. The whole farcical incident resulted in a lengthy investigation costing £144,000, the arrest and suspension of several police officers and seven findings of police misconduct.

The electorate might throw their hands up in disgust at their politicians' gross displays of greed, lasciviousness and abuse of power, but they can only end up banging their heads against the wall in the face of the excuses, the denials, and the increasing daftness of ensuing evasive action and cover-ups.

As Napoleon supposedly remarked, 'In politics, stupidity is not a handicap.'

THE PARADISE
OF SLEAZE

Hand-tinted print
56 × 76cm (22 × 30in)

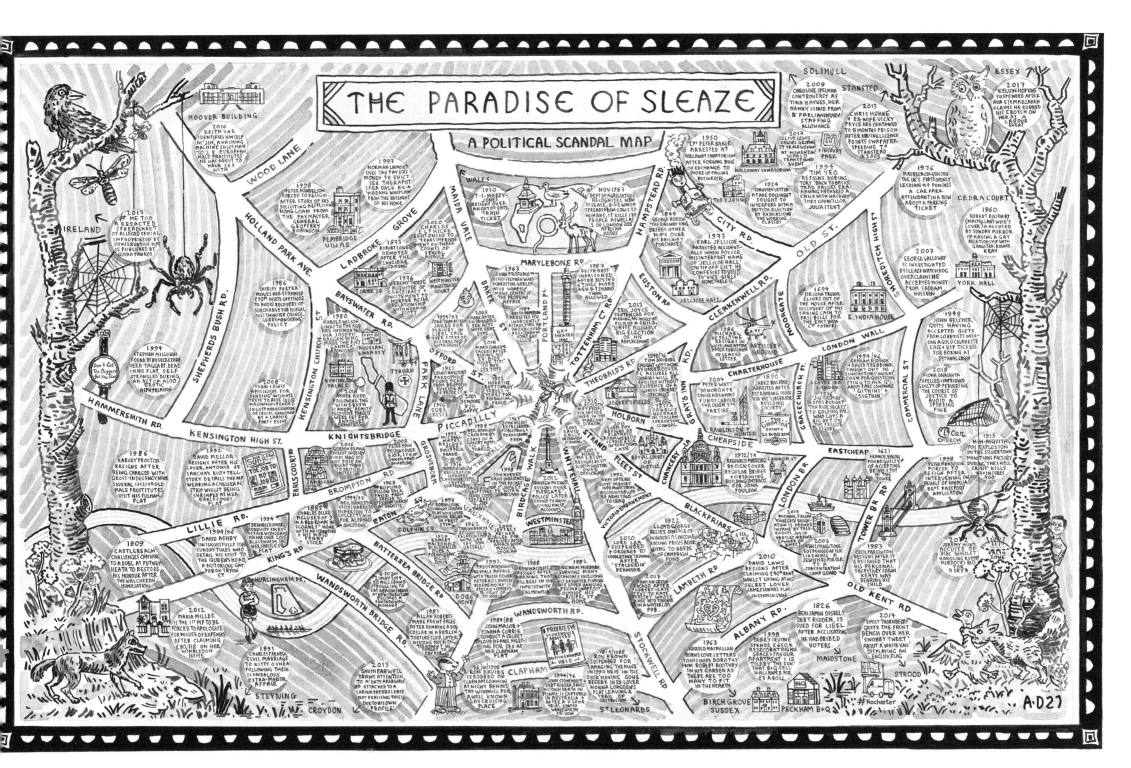

THE GREAT BRITISH BEAST CHASE

Just as the political life of a country can amalgamate all manner of beliefs and opinions via the unifying conduit of 'politics' and its attendant measures and mechanisms, the far deeper 'folk' roots of the disparate cultures that coalesce as nations are similarly underpinned by the alternative 'certainties' of 'myth' and 'superstition'.

When presented with 'The Great British Beast Chase' map of the UK, a group of visiting Americans were at first delighted by all the strange and wonderfully creepy hybrid creatures and characters described. That was until, after a long discussion on the subject of 'fairies', 'piskies' and the other phenomena with which they were familiar through the works of Lewis Carroll and J.K. Rowling, the full absurdity dawned on the tourists that their cousins back in the 'Olde Land' actually believed that such 'critters' were real living things flitting around the periphery and causing mischief in the daily lives of the Brits.

The presence of the likes of the banshee, the wood-wose, Jack Frost and leprechauns appears to be stronger in those regions of the British Isles whose origins are more rooted in the story of pre-Christian and early Christian settlements, as well as in communities who have a long and continuing lineage and involvement with particular trades and activities such as fishing, mining and animal husbandry. A good friend at art school, an Irish girl who had pretty much never left her family farm under the Mountains of Mourne, brought with her to the big city not only the charmingly dewy habit of greeting everyone she passed in the street, but also a whole array of strange ticks and foibles conditioned by the presence and activities of 'the little people'. For her, both the random pedestrian and the wee grey men in pointy hats (or whoever they were) had to be acknowledged and kept happy, content and appeased, lest they wreak all kinds of havoc with their pesky ways. Visiting her shortly before our dissertations had to be handed in, I found her taping orderly sheets of typed-up text to all the walls of her digs. Apparently she kept finding that her writings when stacked as a pile of papers on the desk were constantly being shuffled out of order by the 'you know who'. All this, despite having left the fellows a bit of apple and blackcurrant pie next to the fireplace.

THE GREAT BRITISH BEAST CHASE

WORK OUT WHO'S WHO!

Become an Expert beast-spotter by matching up the names to the beasts, and you will never muddle your 'mermaids' with your 'melusines' or call a Boggart a 'Buggane' again! Just fill the numbers in the circles when you know who's who!

① ABARIMON : He greets everyone with the back of his head, as his feet don't face forwards but backwards instead.

② ADAR LIWCH GWIN : This extremely tall bird doesn't twitter or squalk, when it opens it's beak, it's in Welsh that it talks.

③ ATOMY : Atomies are teeny suprisingly so, but they're good with a scythe or a rake or a hoe.

④ AFANC : It's as loyal & as kind as a Golden Retriever but has spines like a lizard & teeth like a beaver.

⑤ BANSHEE : They appear in your room in a ghostly white cloud, then they howl & they 'howl' incredibly loud.

⑥ BISHOP FISH : He has legs and a face like an ordinary 'bishy' but, the rest of his body is totally fishy.

⑦ BONNACON : From the Bonnacon's bum I would swiftly retire, unless your desire is to end up on fire!

⑧ BLACK SHUCK : They say that this hound has one glowing red eye & they say, if you see it, by morning ... you'll die!

⑨ BOGGART : A Boggart is nasty and naughty and trixy and his hat, belt and boots,,, he stole from a Pixie.

⑩ BROWNIE : These beasts are like Boggarts but softer & funny, their mouths are more smiley, their hats 'more' pom-pommy'.

⑪ BUGGANE : He's not so big, he can dance a good Jig and his nose sticks up like a piggly-wig.

⑫ CANVEY ISLAND MONSTER : His big buggy eyes make him look rather charmless, at first he seemed scary, but turned out, he was 'armless.

⑬ CATOBLEPAS : It's body and tail are all horrid and scaly but the hair on its head is quite silky & wavy.

⑭ CERICOPITHICUS : There were once 2 twin apes & they clung to their mummy, one on her back and one on her tummy.

⑮ DULLAHAN : When they don't have a ball in the land of the dead, they ask nicely to borrow poor Dullahan's head.

⑯ HUNTING DWARVES : Dwarves don't ride horses when hunting, but goats, and they don't hunt for tigers but geeses and stoats.

⑰ DERBYSHIRE FAIRY : Faries exist! Tell everyone! I've found a fairy skeleton!

⑱ FAIRIES : Maybe Faries travel in threes? the more you look, the more you sees.

⑲ HAG : 'what's in your bag, wretched hag?' 'golden coins and a gossip mag!'

⑳ INDUS WORM : It ate all the Lambton family up, so it says in the Lambton museum, it found them by using it's keen sense of smell, with no eyes, it was quite hard to see-'em'.

㉑ JACK IN IRONS : Often times, Jack in Irons, finds that he is in two minds.

㉒ JACK FROST : "We can tell what you had, for your breakfast Jack Frost" the childrens shouts made him quite deaf, 'How can you tell?' Jack Frost replied, 'we know from your frosty breath'.

㉓ LEPRECHAUN : The leprechaun, or so we're told, sits atop a crock of gold, but if you care to ask for some, you'll have to rummage neath his bum.

㉔ LIAM HIGYN Y DWR : If being a frog-bat-lizard, wasn't already quite enough trouble for most of the folk who meet this odd bod his name is unpronouncable.

㉕ LOCH NESS MONSTER : When she stuck her head up out the water, a local photographer caught her, Nessie never came back, for a less blurry snap, though he waited an hour and a quarter.

㉖ MELUSINE : The Melusine lives in the brine and twixt two tails will you entwine, and very soon she will be mine, I've got her on my fishing line.

㉗ NUCKELAVEE : The Nuckelavee lives in the sea, part fish & part horse and part 'fiddle-di-dee.

㉘ PANOTTI : The best part of having giant ears, is of course the exceptional hearing, the worst is the Jewellry you're given as gifts 'bicycle wheels' for earrings.

㉙ PICTISH SWIMMING BEAST : It could be a snorkel on top of his head, or very long curls of his hair, but no! as the 'Pictish beast likes 'the world service' he's planted his aerial there.

㉚ PUCK : "I am sent with broom before, to sweep the dust behind the door."

㉛ MERROW : She watched me as she combed her hair, I felt the mermaids 'Finny-Stare'.

㉜ SPRIGGAN : The guardian of the graveyard, the scary twiggy Spriggan, she guards against the grave-robbers a diggy-diggy-diggin.

㉝ VEGETABLE LAMB : I really like this veggie-beast, I like it quite a lot Just chop it, dice it, boil it up 'hey-presto! quick hot pot.

㉞ WATER HORSE : The water horse is back again from galloping the Spanish 'mane'.

㉟ WOOD WOSE : The Woses Wose by the Woodwose were woses wose on wood I Wose my woses the woodwose way as the woodwose knows I would.

Adam Dant 2009

©Ab09

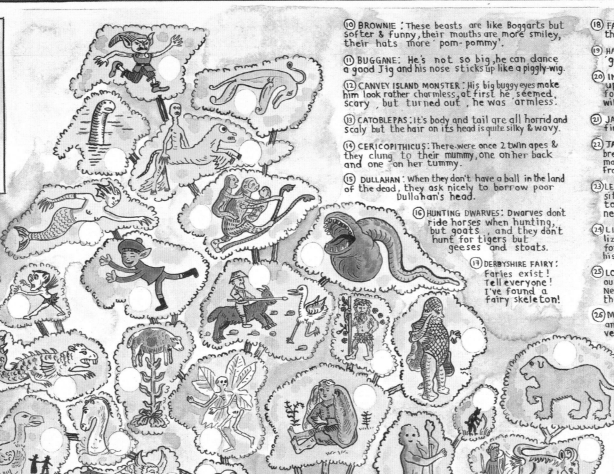

THE MEETING OF THE OLD & THE NEW EAST END IN REDCHURCH ST

In recent years, Redchurch Street has become the conduit through which the culture of the New East End has been channelled into the Old East End, as the street that was once part of the infamous Old Nichol slum has been transformed into London's most fashionable destination. At one end is Shoreditch, lined with new media enterprises and expensive bars, while at the other end is Brick Lane, with its street markets, leather shops and beigel bakeries. And in the middle is Adam Dant, the last artist left on Redchurch Street, living and working in the midst of the hullabaloo.

Celebrated as a cartographer extraordinaire, Adam took his satirical pen in hand to create this epic social panorama of 'The Meeting of the Old & the New East End in Redchurch St', populated with hundreds of characters, both real and mythological, that compose the identity of this notorious thoroughfare. At the far end, drunken Foxtons estate agents characterized by their pointy shoes and spiky haircuts collapse in a drunken heap while leaving The White Horse strip pub. Ken Livingstone is about to run them over by driving a bendy bus round the corner as Boris Johnson falls off the open platform of a passing Routemaster. All this drama, yet you are merely on the threshold of Redchurch Street.

Meanwhile, among those representing the Old East End, you may recognize Richard Lee, the bicycle-parts seller whose family have been trading on Sclater Street since 1880, and the young Charlie Burns, the legendary waste-paper merchant who died in 2012 at 96, portrayed here at seven years old, put in a halter by his father to pull the waste-paper cart round the City. Elsewhere in this extravagant fantasy (eerily not too far from the reality) the iconography of the Old and the New East End appear to have become mixed up as members of Shoreditch House have relocated from their rooftop swimming pool to a flooded hole in the road and a pop-up brothel opens for business nearby. Among the mayhem unleashed in this tiny street by the surreal culture clash between flashy new money and long-term poverty, spot Terence Conran, Keira Knightley, Bud Flanagan, Pearlies behaving badly, a pack of dogs from Hoxton and urban foxes on the prowl.

'We are presented with a plastic version of the authentic, here at the City fringe,' Adam confided to me in a discreet whisper as we walked together down the street in question. 'In Redchurch Street, behind this scruffy fascia of poverty, people on laptops are designing apps.'

For local cognoscenti, Adam's drawing is a chance to test your people-spotting skills, while for the rest of us it is a welcome opportunity to chuckle at human folly.

© The Gentle Author

OPERATION OWL CLUB

Operation Owl Club was the name for the formation of 'The Reading Children's Police Force' (Reading being the Berkshire town, as opposed to the edifying activity) founded as part of the local council's new signposting scheme.

Compelled to incorporate a contribution of an artist in their scheme, the council had allocated a small bit of each information panel for such, but being at the very bottom of each sign these could only be viewed comfortably by viewers who were less than four feet tall. In other words, eight-year-olds.

Every schoolchild in Reading was presented with an Operation Owl Club sourcebook that explained the origins of 'The Polis', civil society and the city state, and offered the opportunity to create new states with new reasons to police the public arena according to the raison d'etre of such new communities and the caprices of the eight-year-olds who invented them.

New societies were imagined based on the importance of certain codes of dress, on outwardly positive demeanour, on the behaviour and control of wildlife in the city centre and a host of other fairly draconian systems of government.

A 40-strong group of children were sent out on a policing operation and stationed in small groups at key junctions throughout Reading town centre, with clear lines of sight to each other so they formed a kind of living surveillance system. The network formed by these 'little watchers' is depicted in map form, thus picturing the city as a place of constant monitoring, albeit in a fashion more benign than mass surveillance for the purpose of social control. Police uniforms were designed and produced and various operational activities logged and analyzed.

Turning the map of the city centre into a grid of interconnected roundels left blank for the Children's Police Force to fill in with pictures of what they saw while on patrol subjects the 'forensic' aspects of daily life, even if they are lived in a grid, to the more friendly and haphazard processes by which citizens 'bimble' through the 'polis' according to the whims and wiles of chance.

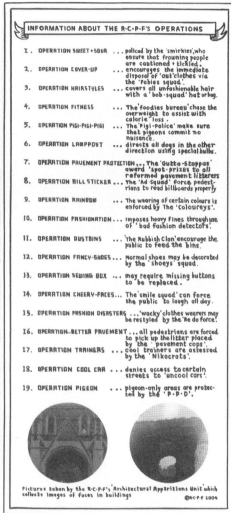

Ephemera from
'Operation Owl Club'.

**OPERATION OWL CLUB,
MAP OF READING
TOWN CENTRE**

Printed pamphlet

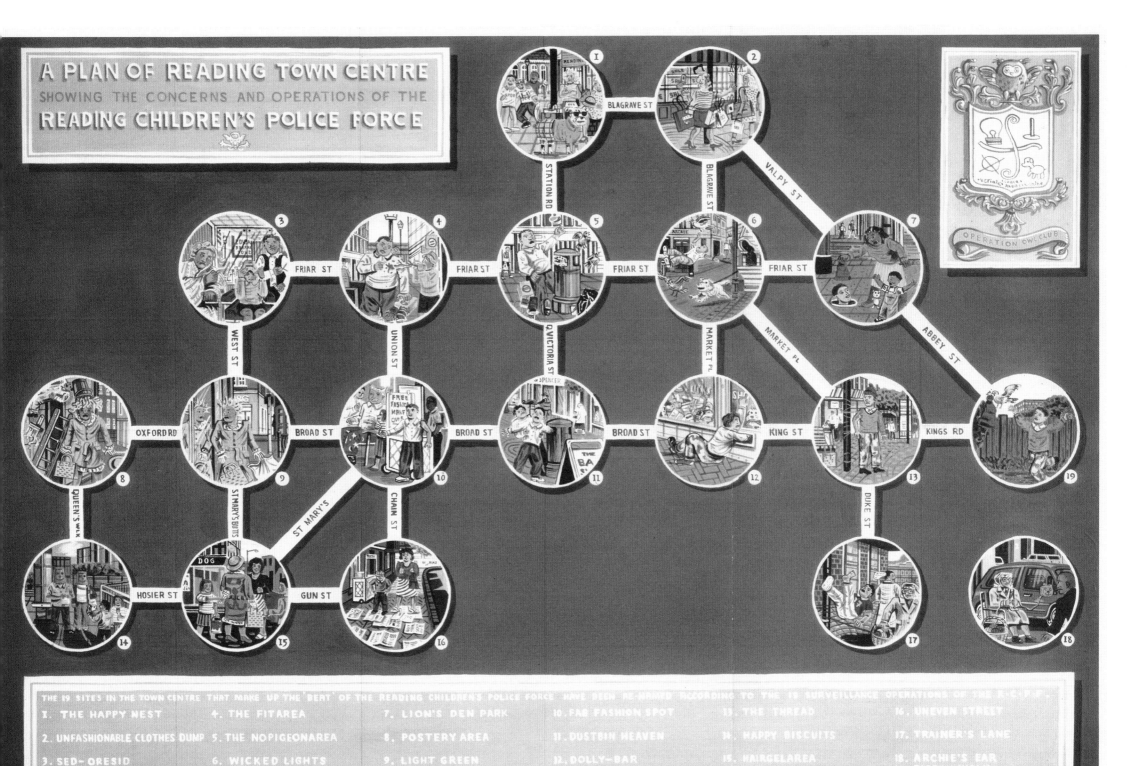

Constructed as if it were a yellowing page torn from the quintessential enlightenment project, Denis Diderot's *Encyclopedie*, this 'Encyclopedy of Ye Age of Enlightenment Citizens and Kings' presents key figures from the golden age of rationalism in their own words, grouped according to particular milieus.

Radiating out from the light of science and understanding, the ordered and schematic nature of the image is, however, framed by depictions of pratfalls, stupid accidents and puerile Three Stooges-style idiocy. Voltaire's pants fall down as he tumbles from atop a bookcase, one of the Montgolfier brothers lands on his head after his hot air balloon breaks free, while other wigged thinkers are struck by tumbling tomes or are thrown to the ground after leaning on a spinning globe.

This is a state of affairs that possibly has more in common with our own age of 'dumbing down', 'unqualified inclusion', 'twitter spats', 'covert social control' and the ubiquitous culture of 'cretinisation' peddled by popular TV shows that would no doubt confidently present this particular 'Encyclopedy' image as a 'back-in-the-day celebrity cake' rather than as a 'taxonomy of the age of reason'.

AN ENCYCLOPEDY
OF AGE OF
ENLIGHTENMENT
CITIZENS AND KINGS

Three-colour lithograph
76 × 76cm (30 × 30in)

STANFORDS WORLD OF COVENT GARDEN

Edward Stanford's Covent Garden store is said to sell the largest collection of maps, globes and maritime charts in the world, having supplied cartography to, among others, Robert Scott, Ernest Shackleton, Florence Nightingale, Amy Johnson and James Bond. Sherlock Holmes pays a visit to buy a map of Dartmoor from Stanford's in Conan Doyle's tale of the frightful Hound of The Baskervilles.

Since 1901, when Stanfords was the only map-maker in London, the shop has been at the heart of Covent Garden life, both geographically, as it is shown at the apex of this map, culturally, and since the store's move from its famous location on Long Acre to 7 Mercer Walk, socially, finding itself in the middle of a newly created pedestrian arena among new restaurants and hangouts.

This anecdotal World of Covent Garden map is thus very much worthy of having Stanfords as a unifying presence for all the topographical, historical and narrative information depicted.

Presumably maps have played a small part in the journeys of those travellers who have Covent Garden as the final stop on their travels. Stanfords map shop is certainly a destination in itself. As with maps that show the holy land or Mecca, this map is dotted with points of interest for the urban pilgrim; places to visit such as The Central Market, Rules restaurant and the Garrick club, as well as recounting odd tales from a realm rich with history and anecdotes of ghosts, crooks, riots, canine actors, clowns, crooners, flower girls, philosophers and map-makers.

STANFORDS WORLD OF
COVENT GARDEN

Hand-tinted print
76 × 112cm (30 × 44in)

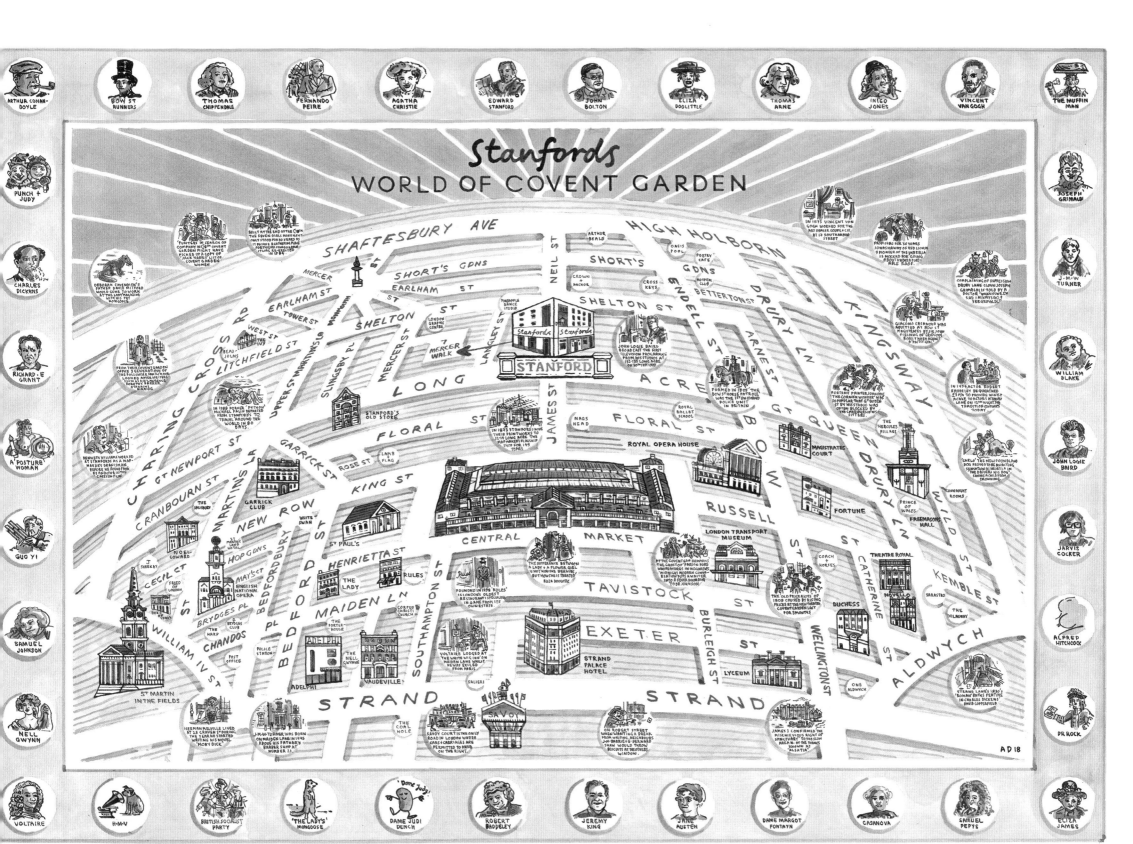

ACKNOWLEDGEMENTS AND CREDITS

The author would like to thank Melissa Brady-Dant, Grace Dant and his walking companion, who is often featured on these maps, Doctor Watson the Scottish terrier, The Gentle Author, Melanie Unwin and the curatorial office of the Parliamentary Art Collection, Hobby Limon and all the staff at TAG Fine Arts, The Chelsea Arts Club.

Limited edition prints of Adam Dant's maps are available from www.tagfinearts.com

Page 12 Unknown, *Almanac for the Year 1734 (The August Portraits of the First Born Sons of our kings)*, 1733; etching and engraving on paper; Waddesdon (National Trust) Bequest of James de Rothschild, 1957; acc. no. 2669.3.17. Photo: Waddesdon Image Library, Mike Fear.

Page 48 Peter Marshall/Alamy Stock Photo.

Page 73 (top right) Eric Dahyot.

Page 98 Private collection, Pierre Haddad.

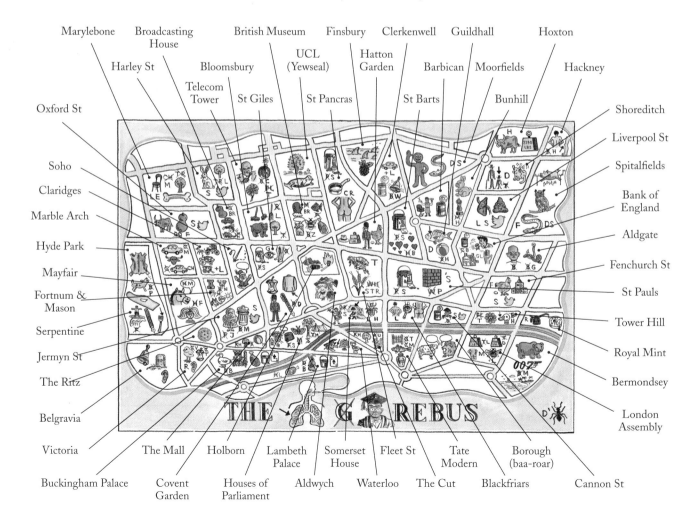